CRITIQUE OF
SCIENTIFIC
REASON

IN MEMORY OF ELLEN ROSER, M.D.

Das erste steht uns frei,
beim zweiten sind wir Knechte.
Goethe

The first step stands free,
with the second we are vassals.

CRITIQUE OF SCIENTIFIC REASON

KURT HÜBNER

TRANSLATED BY
PAUL R. DIXON, JR., &
HOLLIS M. DIXON

THE UNIVERSITY OF CHICAGO PRESS·CHICAGO & LONDON

Kurt Hübner is Professor Ordinarius of the Philosophisches Seminar of the University of Kiel.

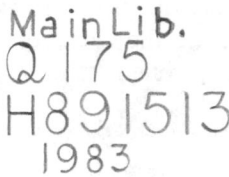
THE UNIVERSITY OF CHICAGO PRESS, CHICAGO 60637
THE UNIVERSITY OF CHICAGO PRESS, LTD., LONDON

© 1983 by The University of Chicago
All rights reserved. Published 1983
Printed in the United States of America
90 89 88 87 86 85 84 83   5 4 3 2 1

Originally published as *Kritik der wissenschaftlichen Vernunft,*
© Verlag Karl Alber GmbH Freiburg/München 1978 (2d ed., 1979).

The University of Chicago Press gratefully acknowledges a subvention from Inter Nationes in partial support of the costs of production of this volume.

LIBRARY OF CONGRESS CATALOGING IN PUBLICATION DATA

Hübner, Kurt, 1921–
  Critique of scientific reason.

  Translation of: Kritik der wissenschaftlichen Vernunft.
  Includes bibliographical references and index.
  1. Science—Philosophy. 2. Quantum theory.
3. Technology—Philosophy. I. Title.
Q175.H891513 1983   501   82-23690
ISBN 0-226-35708-2
ISBN 0-226-35709-0 (pbk.)

CONTENTS

PREFACE

Today many people believe that truth and knowledge in the proper sense exist only in science, and therefore that all other aspects of existence must gradually be brought under the control of science. Furthermore, there is a widespread opinion that the quality of human life depends essentially upon scientific enlightenment. Hence, perhaps now more than ever, extrascientific realms, such as art, religion, and myth, give rise to embarrassment: How can these realms still be taken seriously and justified? But the opposing view exists as well. In particular, the highly questionable nature of technological progress (air and water pollution, overpopulation, etc.) has led to the propagation of an irrational animosity toward science. Apparently, neither of the above positions adequately grasps what science properly is, what truth, experience, and knowledge mean in it, what science via its central notions is and is not capable of achieving. And all of this could equally well be said for technology.

The following investigation is intended to contribute to the resolution of these problems. Accordingly, new insights will emerge here which will shed new light on the above-mentioned extrascientific realms as well.

It has not been my intention to treat the subject exhaustively (if this is even possible) and to come to terms in a comprehensive manner with all the contemporary literature that has been produced concerning what is to be touched upon here. In general I have limited myself to the most essential themes, in order to present the often unfamiliar fundamental thoughts all the more clearly and in a manner which might facilitate their understanding. In addition, because the subject matter is of an immediate and highly pertinent nature for the modern world in general, I have not directed myself merely to specialists in the theory of science but rather to a wider audience. For some readers a few of the chapters may prove difficult; but this will in no way make an understanding of the entire work impossible, especially as the most important chapters demand no special prior knowledge and, further, are so constructed that they may be read as independent units. In particular, chapters 1, 3, 4, 8, 11, 13, 14, and 15 belong to this latter category.

Finally, there is one more point which should be emphasized to avoid misunderstandings: The present treatise is devoted to the sciences only insofar as these take the *form of empirical theories,* a form they have taken on in the course of modern times. However, it should also be stressed that the pressing present-day problems alluded to above are only related to science when understood in this form.

I thank those people who worked with me at the University of Kiel, Dr. Deppert, Dr. Fiebig, and Mr. Sell, for their numerous suggestions and for their reading of the manuscript. In addition, as several chapters of this work originated while I was teaching in Berlin, I would also like to thank my former colleagues, Professor Lenk, Professor Rapp, and Professor Gebauer, for the many stimulating suggestions they gave me.

Kiel—31 December 1977

TRANS. NOTE: The numbers in the margins refer to the page numbers of the second edition of *Kritik der wissenschaftlichen Vernunft*, published in 1979. The numbered (end)notes are Hübner's original notes, translated from the German; the lettered (foot)notes are the translators'.

PART 1 THEORY OF THE NATURAL SCIENCES

1

HISTORICAL INTRODUCTION TO THE QUESTION OF THE FOUNDATIONᵃ AND VALIDITY OF THE NATURAL SCIENCES, THE NUMINOUS,ᵇ AND ART

When we turn the key in the ignition switch of our car, we expect the motor to start; in the dark when we flip a light switch, we expect the light to come on; the movement of the stars can be predictably calculated; chemical reactions are constantly produced by the same means. Our entire life in the world of industry is filled out by a closely woven mesh of technical activities which constantly entail expectations that are constantly being fulfilled. At the base of all this, however, lie physics and the laws of physics. For this reason Lenin asserted that praxis proves the truth of physics.[1] 19

Physics, however, not only contains laws which we are constantly employing; it also gives a definite interpretation to manifold appearances both in the physics laboratory and in everyday life. A light bulb lights up and we say there is a flow of current. We observe the rising tide and say the moon exerts a gravitational force. When we see light, we say here are electromagnetic waves. We hear a radio and say a voice reaches us from the air. When we observe a trace on a plate from a cloud chamber,

a. The *foot*notes are the translators' notes; the endnotes are Hübner's.

Begründung, translated here as "foundation," refers to a complex idea which is difficult to render into English. The term is a substantive form of the verb *begründen,* which refers to the activity of founding or establishing something (in this case, science). This active meaning is carried over into the substantive form in German. Accordingly, the reader is advised to remember that Hübner is not merely dealing with the static concept of the foundation of science (or that which "substantiates"—literally "underlies" and "supports"), but moreover with the act of founding itself. Hübner is concerned here with the conditions inherent in such an act.

In addition the term has a second, parallel meaning, relating to the substantiation process or proof of a given idea or theory. Hence it also means to offer evidence or reasons in support of a theory.

All of this must be held in mind if the term is to be adequately understood.

b. *Numinose*—The word "numinous" or "numen" is used primarily in theology with reference to divine power. In its Latin root, "numen" designates a sign or effect of a god or a divine being. As Hübner indicates (cf. p. 9) the term is not applied merely to Christianity or any other specific religion; its meaning has to do with the fundamental experience in which all religion is grounded.

we say this is the picture of the path of a particle, and we name the particle an "electron." In all of these cases physical theories have infiltrated the everyday language of modern man as if they were self-evident. These theories are then also held to be true. And yet neither the existence of physical laws nor the truth of physical theories is, as constantly suggested to us, self-evident; rather, each is something questionable.

The present chapter should serve to introduce the questionable nature of physical laws and theories. This will be done by discussing three classical and pertinent examples of the manner in which these laws and theories have been treated. These are: Hume, representing critical empiricism; Kant, representing transcendentalism; and Reichenbach, who (in this case) represents operationism. In this manner I will also show how the problematic of the numinous and of art comes to the fore in terms of these historical examples.

1.1 The Question of Foundation for the Natural Sciences in Hume's Critical Empiricism, Kant's Transcendentalism, and Reichenbach's Operationism

We begin with Hume, and in connection with this let us take the law of falling bodies as an example:

$$d = \frac{g}{2} t^2 \,.$$

If the time at which the body began to fall is known, then by means of this law we can predict the distance it will have covered at a later time. In supposing that physical laws exist, the claim is made that these laws express a universal constituent structure (*Verfassung*) of nature, that nature is indeed truly construed according to these laws. Hence these laws are *always* supposed to be valid, and this obviously means in the future as well, since only if this is the case are they indeed laws. But the experience of having such laws and constantly achieving success through the acceptance of such laws is something which belongs only to the past. Accordingly, all we can ever say at most is that hitherto our expectations, based on these laws, have always been confirmed. But then, by what right do we infer in such cases from the past to the future and assert that these laws will always be valid, because they are *universal* laws of nature? The experiences drawn from practice in no way give us the right to make such an inference. To call upon these experiences, as can easily be seen, would lead to circular reasoning. The argument would have to run as follows—and in fact such is the manner in which it is popularly argued: Hitherto, that is, in the past, we have had practical success in our inferences from the past to the future; therefore, in the future we will also have success by means of such inferences; and thereby the form of reasoning is justified. However, this founding already makes use of what it desires to found; namely, it makes use of an inference from the past to the future, since indeed it simply infers from past practical successes to future ones. Recourse to pure logic would also be of no help here, since the constancy of rules in nature, which must be presupposed by all inferences of the type referred to here, is not a concept of logic. Logic, in its empty and formal universality, teaches us nothing about particular characteristics of nature, and hence nothing about their constancy. Thus the result is summed up in the following: Neither experience, which is always past, nor pure logic can ever prove the existence of physical laws that would be valid for all times. This was Hume's fundamental insight.

In this way it becomes evident that neither the existence nor the content 22 of natural laws is in any way an empirically given fact. We do not simply find these laws in nature. Rather, we apparently bring them to nature in a certain sense: we impose them on nature. Whenever, over and above this, we suppose that in spite of this they still exist as intrinsic to nature itself—that there must be, so to speak, a preestablished harmony between what we bring to nature and what in truth exists in it—then we must clearly come to recognize that such a supposition cannot be founded. It is rather the expression of a belief.

But then by what right do we transfer laws to nature? Hume simply referred their employment back to habit, which orients itself by means of constant repetition of an orderly series of consecutive events. In so doing he completely abolished the question of right just mentioned. However, this view is untenable. In the first place, a habitual orientation can only be spoken of in the case of fairly simple laws, laws which have a form something like: "When you reach into the fire, you burn yourself." But we also trust nonetheless in laws which are far removed from the realm of continual experience and habit, as for example the laws which determine the orbits of comets. Second, the acceptance of the kinds of laws which underlie science cannot be grounded on something as fickle and subjective as human habituation. For this, science needs rational grounds.

Therefore, we must ask, by what right are physical laws presupposed if they are not given in experience, and thus if their existence is in no way guaranteed?

Kant begins by assuming that we must necessarily think the manifold, scattered representations which fill out our consciousness as standing in 23 a possible, thoroughly continuous connection. For only if they do stand in such a possible connection can they belong to the unity of an ego consciousness (*Ichbewußtseins*). And so within our consciousness the representation of a universal and integrated world horizon, within which we arrange and order everything, resonates in a continually intuitive and more or less thematic manner. However, these interconnections are not given to us actually, nor are they continuously given in experience. They must only be thought as fundamentally possible by an ego that understands itself as unity; and accordingly they must be presupposed a priori. It then becomes Kant's task to seek out these a priori presupposed interconnections by means of which, as he thought, consciousness constructs itself as a unity. In this way he arrives at the conclusion that there must belong to these interconnections, among other things, the connection of the representations of events according to the principle of causality. If we leave aside certain insignificant problems, this principle can briefly be stated as follows: For every event there is a causal explanation such that it must be thought as arising out of previous events in accordance with a universal rule. In addition, this principle appears to be the condition for the fact that representations of events are given to us as objective in any sense at all. And this is held to be the case because a representation of an event is objective, and not subjectively capricious, according to Kant, only if it "stands under a rule, which distinguishes it from every other apprehension and necessitates some one particular mode of connection of the manifold."[2] And the causal principle is, for example, such a rule. Only when we think of an event as arising in accordance with a causal rule do we view it as objectively true and not as originating from our arbitrary choice. The a priori form of the causal principle is therefore, as Kant says, the condition not only for the possibility of the unity of consciousness, but also for the fact that experience of objects is possible in any sense at all.

Thus, if in the past we have discovered a rule like the law of falling bodies, we henceforth have a right to expect it to be valid in the future as well; for this rule is then indeed nothing but a particular case of an a priori valid assertion of the causal principle that all events must be thought by necessity as originating in accordance with invariable rules and laws.

So we find this to be the answer which arises out of Kant's transcendental idealism to the question: "By what right can physical laws be presupposed a priori if they are not given to us empirically?"

In opposition to this, Reichenbach's operationism answers the above question in the following manner: If the goal of science is to make prognoses and to master nature, then it must presuppose that natural occurrences take place in accordance with constant rules and laws. The existence of such laws cannot be proved in a purely empirical manner; but since there is only one methodological way, if indeed any way at all, to reach the desired end of prognosis—namely, by presupposing laws—we must take this road, even if we cannot know in advance whether our efforts will be in vain.

"A blind man," writes Reichenbach, "who has become lost in the mountains, feels out a path with his cane. He knows not where the path leads, nor whether it might take him so close to a precipice that he may stumble over. And even so he will follow the path, going ever further, feeling his way along step by step with his cane. For if there is any possibility at all for him to find his way out of the wild craggy region, it is by feeling his way along this path. As blind men we stand before the future; however, we feel out a path, and we know that if we can find any way at all through the future, it will come to pass by feeling our way along this path."[3] Reichenbach means to say the following with this analogy: Whoever involves himself in physics and wants to master nature must in his method presuppose a priori physical laws and the causal principle. But in this there is no assertion made concerning the existence of such laws. Reichenbach's formulation can also be carried over into everyday life. Why do we incessantly presuppose natural laws for the most insignificant activities? Precisely because we want to act, and this desire to act, when seen from a rational point of view, entails the presupposition of such laws.

1.2 Comparison of the Foundations
of Transcendentalism and Operationism

A comparison of Kant's transcendentalism and Reichenbach's operationism yields the following results: The *Critique of Pure Reason* has the comprehensive goal of demonstrating that the foundations of physics—for example, the categories of causality and reciprocity, etc.—present us with the necessary a priori framework within which objects can be given to a unified ego consciousness in any sense at all, and thus that framework which makes possible any experience at all. Hence, for Kant, there is only a difference in degree between the physical-theoretical manner of viewing things and that of everyday life: with regard to its foundations, physics merely explicates what every ego consciousness unconsciously posits in advance (*voraussetzt*) a priori.

Therefore, for Kant, physics remains, at least in terms of its form, the single justifiable way of viewing the external world. Obviously, Kant went even further in his later writings. In the *Metaphysical First Foundations of the Natural Sciences (Metaphysische Anfangsgründe der Naturwissenschaften)* and in his *Opus Postumum* a great deal of the *content* of Newtonian physics is deduced a priori as well.

In opposition to this, the operationist standpoint states that physics is neither true nor false; rather, it rests upon a priori precepts (*Festsetzungen*)[c] and ideal constructions which are imposed on nature only to

25

26

c. *Festsetzung(en)*, translated here as "precept(s)," is a term which receives an elaborate treatment in chapter 4 (cf. pp. 42–47). *Eine Festsetzung* is basically an established or firmly held position underlying the sciences, hence, an "established" or "accepted" precept. (In what follows, these adjectives will sometimes accompany the noun.) Hübner's position, as elaborated in chapter 4, is that certain "precepts" are necessarily presupposed by the sciences.

the end of creating a schema for its mastery. But no claim is made that these constructions delineate the constituent structure of nature itself. This can perhaps be compared with a network of coordinates we might superimpose on the sphere of the Earth in order to orient ourselves upon it. This network is also an ideal construction, but it is not a property of the Earth.

Hence, what transcendental philosophy and operationism have in common resides in the fact that both have given up the classical naive ego-object relation according to which an object *in itself* stands oppposed to the ego and from which object the ego then acquires an understanding (*Kenntnis*) by means of experience. Further, both transcendental philosophy and operationism teach that the ego itself produces the object in a certain respect. Therefore in both cases we have something aprioristic, insofar as "a priori" means simply: not given through experience but rather from ourselves. The difference, however, consists in the fact that Kant holds the manner of this production to be a priori necessary and unalterable, while operationism derives it purely methodologically from the goal of viewing nature in terms of its mastery. Accordingly, the Kantian "a priori" could be called necessary; that of operationism on the other hand could be called contingent or arbitrary. For Kant physics is the only possible genuine object construction; for operationism, by contrast, physics rests upon a particular decision. Hence, from the perspective of transcendental philosophy, the history of the genesis and development of physics, with all of its incalculable historical consequences, as indicated in industry, the creation of the atom bomb and rockets to the moon, etc., must all be viewed as a process by which reason first makes clear the manner in which it objectively constitutes objects in any sense at all. On the other hand, from the perspective of operationism, the genesis and development of physics rests upon an act of the will, a will to power over nature—just as Bacon and Hobbes, these first men of the technological age, had already sensed.

1.3 The Question of Foundation for the Numinous Realm and for the Object of Art in Terms of Transcendentalism and Operationism

Thus we come to the third section of this chapter, which is intended to indicate how the problematic of extrascientific objects—namely, the problematic of the numinous and of art—develops out of the context of the given historical examples.

For Kant physics is the proper way of viewing the external world. In this sense, then, he is as modern as anyone, when we consider matters in terms of the present-day situation as I have already depicted it. In the technological age man is constantly and imposingly confronted with the object as understood in terms of physics. His entire world is filled with daily technical functions; his language is permeated with physics. He even sees the objects of everyday life through the eyes of physics. Crystal, precious stones, the sea, the sun, the wind—all of this is only explicable to him in the final analysis on the basis of physics, and as such only as material substance. As the popular notion runs, we are in truth dealing with clumps of atoms and elementary particles—and nothing more. Whether this notion is indeed true, half true, or false, we can in any case state that it mirrors the constant and powerful suggestion dominating and subconsciously determining modern man, incessantly besieged as he is on all sides by technology and technological creations. Moreover, one could not succeed in making the foundations of the technological age more invincible than by demonstrating, as Kant sought to do, that physics is a priori necessary. Thus, despite the centuries separating us from him, Kant may be considered a valid exponent of our age.

Now, although on the one hand Kant announced the omnipotence of physics, he sought nevertheless at the same time to limit it once again with the help of some idealistic sleight of hand. Physics, according to him, is supposed to be valid only for appearance and not for the realm of things in themselves. But it is precisely in this that the decisive consequences of his philosophy announce themselves: Physics—he actually says "knowledge," but for him this is the same thing—should be "limited (*aufgehoben*) and room created for faith."[d] But what then is the end effect of this? A rather weakly grounded postulate of a world policeman (*Weltpolizisten*), entrusted with the function of looking after the moral world order. This is the divinity that Kant offers us. In this case, then, an apple orchard was promised and only an apple was given. 29

However, from the Kantian point of view the numinous is impossible. And this word does not indicate merely a Christian conception, but rather a universal religious category. The numinous is the holy, that which, as Rudolf Otto has depicted it, makes man tremble before the "mysterium tremendum" and at the same time draws him along in the wake of its "mysterium fascinosum."[4] The numinous is the divine that appears to us in space and time; and precisely for this reason it is the miracle. But it is precisely the miracle which must be impossible for Kant, since it occurs in the realm of appearances, for which physics alone is a competent judge. For this reason, then, there is a retreat into the intelligible world of the in itself, in virtue of which the abstract world policeman is postulated. What Hume said about miracles can be considered just as characteristic of Kant. Hume taught that whenever a miracle is thought to have occurred, one must consider which is more probable according to the scientifically and extrascientifically recognized laws of nature and the human psyche:

d. This is a paraphrase of the famous line from Kant's *Critique of Pure Reason* (B xxx): "Ich mußte also das *Wissen* aufheben, um zum *Glauben* Platz zu bekommen." (Thus I had to limit *knowledge* in order to make room for *faith*.) The crucial word in this passage is *aufheben*, which here has its primary meaning of "to cancel" or "to limit" (partially cancel).

that the miracle has taken place or that it is a deception. And only if the deception is more miraculous than the miracle can the miracle be accepted as true. However, with respect to the laws in question, this will never prove the case. From such a point of view, then, the numinous is impossible. Obviously, the fact that Hume scarcely had a right to such an argument, since, in opposition to Kant, he doubted the possibility of grounding natural laws, is another matter altogether.

30   Furthermore, viewed in this light the object of art (*Gegenstand der Kunst*) also becomes impossible. I would like to use the example of the visual arts (*bildenden Kunst*) to make this point. What is the "object" of art? Here I do not mean the subject or theme of a painting, but rather the totality of the experience the painting presents to us. And this obviously covers the so-called objectless art as well. This object of visual art is not the same as that of science in general or that of physics in particular. It is, so to speak, immune from an objectivity construed according to scientific laws.

All theories of art, from those of the ancients up to but excluding that of Kant, proceeded on the basis of the object of art; and in doing so they variously related this object to the Platonic "Idea" or the Aristotelian concept of form.

It is indeed symptomatic that Kant, by mirroring the tendency of the age to make physics the criterion of judgment for objects, divests the artistic object of its meaning. He does not lay the emphasis of his theory of art on the object, but rather on the effects which the artwork incites in the viewer. These effects, according to the Kantian notion, consist in disinterested pleasures (*interesselosen Wohlgefallen*) and the beneficent, free, and harmonious play of the powers of cognition (*Erkenntniskräfte*). The object of art comes into view only insofar as it is supposed to have the universal form of purposiveness (*Zweckmäßigkeit*). However, for Kant purposiveness is not an objective constituent structure of nature, but rather only a subjective principle of judgment. According to this notion it would thus be appropriate, although somewhat exaggerated, to say that the prototype of the artwork is the wallpaper pattern.

31   But when Kant thus relocates the aesthetic experience on the side of subjectivity, it is merely the flip side of the fact that he must pronounce the object of art—that is, the totality of the experience presented in the artwork—to be impossible; and this is the case because what is possible is only that which can be encountered by us according to (in the strict sense, scientifically recognized) empirical or a priori laws. Everything else he leaves to the realm of subjectivity, of fiction, which in the final analysis is insignificant. But when the possibility of the object of art is thus negated, art itself and the aesthetic experience are also negated; for this experience is only made possible by the fact that we claim the right to consider the object of art—whether in terms of the process of creation

or in its contemplation—as something objective. The artwork derives all of its enchanting power, all of its meaning, from the claim that its object is valid in some way or other, that it is a possible interpretation of the real world. Thus Kant sealed off the sole spring from which the numinous and art can draw forth life-giving water.

In contrast to transcendental philosophy, operationism holds open the possibility of art and the numinous; but it is also unable to ground this possibility. According to operationism, the objective sense of physics originates in the fact that—and here it is initially in agreement with Kant—we bring a synthetic unity according to rules and laws into the manifold of perceptions in an a priori manner. But whereas for Kant this is the necessary form of any object construction at all, according to the formulation of operationism it is determined by means of practical ends, and therefore is not necessarily binding. Hence, formally viewed, the object of art originates similarly in that the artist, to paraphrase Kant, introduces 32 a "synthetic unity according to rules" into the manifold of perception. Every artwork has its internal stylistic and structural laws which serve to bind together the realm of the manifold through the introduction of order and form, even if these are completely different from those of physics. In addition, the artistic synthesis, when viewed from the standpoint of the theory of knowledge, is something a priori, namely, a creation (*Schöpfung*). Now, since the form of object construction belonging to physics, from the viewpoint of operationism, enjoys no priority over any other form, it can no longer in principle contradict that which belongs to the artistic object. And the same holds with respect to the numinous, since a physical law is neither true nor false for operationism, but rather only an ideal construction methodologically created in accordance with a specific purpose; hence it poses absolutely no fundamental objection to the numinous.

However, as we have stated, neither the artwork nor the numinous experience can be founded in such a manner: For even if the right to creations other than those of physics can in principle no longer be blocked within the framework of operationism, nevertheless nothing has been decided here about the nature of that which is to lead us to a recognition of what actually imparts objective validity to such different creations.

Thus the question concerning the validity of physics—and, as physics is a kind of fundamental science of nature, the question concerning the validity of natural science in general—has indeed been brought into sharp relief through the discussion of the historical examples introduced here. Further, we now recognize how this question is necessarily related to that concerned with the object of art and the experience of the numinous. Moreover, the answers offered to us by Hume, Kant, and Reichenbach can no longer be considered satisfactory today. What motivated them in 33 giving their answers motivates us today no less urgently. In our present situation, we are more profoundly caught up in the world of physics and technology—a world which on the one hand fascinates us and on the other has alienated us even further from the numinous and art. But in the intervening years, a wealth of new insights has arisen concerning this matter; this forces us to direct our attention onward, as in the following chapters, to new and different paths.

2

A CASE STUDY: THE FOUNDATION AND VALIDITY OF THE CAUSAL PRINCIPLE IN QUANTUM MECHANICS

Having given the foregoing historical introduction, I now wish to turn to   34
a particular case study before continuing with the systematic development
of the theme of this book in the next chapter. In so doing I will prepare
the way for the more abstract and universal treatment to follow.

The causal principle has always been one of the most important principles of the natural sciences; as such it has also been in the foreground
of the philosophic interest in quantum mechanics. Let us therefore ask:
How is the causal principle expressed in quantum mechanics? Is the causal
principle valid in quantum mechanics?

2.1 Limitation of the Applicability
of the Causal Principle in Quantum Mechanics

Heisenberg expressed the principle in "precise formulation" as follows:
"If we know the present exactly, then we can calculate the future."[1]

Now, according to his understanding of this formulation, it is "not the
consequent, but rather the premise which is false. In principle we *cannot*
come to know the present in all of its determinant elements."[2] The reason
for this unknowability is the uncertainty relation of quantum mechanics.[a]   35
This principle asserts that the only thing we can ever measure exactly is
either the location or the momentum of a particle, but not both at the
same time. (When I speak of the uncertainty relation in what follows, I
always mean it in this form.) Now, if as a result of quantum mechanics
the "premise" of the causal principle becomes false but at the same time
all experiments are subject to quantum mechanics, then it follows from
this for Heisenberg that "the invalidity of the causal principle has been
definitively established."[3] The apostles of the theory of "acausality" have
supported their thesis ever since, on the basis of this remark coming from
such renowned lips.

a. *Die Unbestimmtheitsrelation*, translated here as "the uncertainty relation," refers to
what is generally translated into English as "the uncertainty principle." Strictly speaking,
however, there is a difference between the "relation" in question and the "principle" which
describes the relation. Further, the term would be more exactly translated as "the indeterminacy relation"; but as the word *Unbestimmtheit* now has a history of being translated in
this context as "uncertainty," I have left it in its customary form.

However, if we take Heisenberg at his word, then we must maintain that logically viewed his assertion is false. The causal principle is expressed by Heisenberg as an if-then proposition. However, according to the rules of logic, such an if-then proposition does not become invalid when its premise (protasis) is false. To the contrary, if the premise is false *in principle* (in principle we cannot know the present exactly), then the if-then relation (here the causal principle) is indeed always true.

To be sure, in this form the causal principle will never be applicable either. For this would only be possible if we actually knew the present exactly, and thus could calculate the future from it. And for Heisenberg this is never the case.

Obviously, then, Heisenberg has interchanged the validity of the causal principle with its applicability; these, however, are two distinct predicates.

Now, it is not at all difficult to find a causal principle that not only is presupposed by quantum mechanics as a guiding rule but also is applicable there. Such a principle can be expressed as follows: For every fundamentally and exactly measurable event (*Ereignis*), there are other events, simultaneous, past, and future, to which it is connected by means of causal laws. I determine the concept "causal law" (*Kausalgesetz*) on the basis of a definition given by Stegmüller,[4] which I have merely abbreviated here with minor alterations: A causal law is a deterministic law of action by contact (action at *vanishing distances—Nahwirkungsgesetz*), expressed by means of mathematical functions differentiated with respect to the time parameter and related to a homogeneous and isotropic space-time continuum. "Causal laws are deterministic" means that they allow for an exact, and not merely probable, prediction. In physics these laws take on the form of the indicated mathematical functions. They are "laws of action by contact" because the velocity of propagation of sequences of events ordered by means of these laws is finite. They relate to "an isotropic space-time continuum" because the direction in which these sequences of events are propagated is not significant.

In this way the meaning of the statement that an event is connected to other events by means of causal laws becomes clear: It means that these other events can be calculated on the basis of this one event or, inversely, that this one event can be calculated from the others.

The concept of event (*Ereignis*) need not be explicitly defined here. Concerning this we might only say that an event is not defined by the fact that it is essentially and exactly measurable. Accordingly, there might also be events that are not exactly measurable. An example of this would be the so-called interphenomena, by which are to be understood events in microphysics that do not enter into a reciprocal relation with other material phenomena, and thus lie between coincidences—something like the path of a particle lying between its point of origin and its collision with a photon. That such events actually exist is not being asserted; rather this example is only intended to point out that the concept of "event" used here does not include the element of exact measurability.

The example of the measurement of the position of a particle will serve to point up not only that quantum mechanics presupposes the causal principle, formulated in the above manner, as a guiding rule, but also that it is applicable there in such a form.

The causal principle (in the above form) is *presupposed* in such a measurement because the following consideration precedes it: If particular quantities are exactly measured (for instance, the wavelengths of light employed in such a measurement, the dimensions of the measurement apparatus, the resultant diffraction pattern, etc.), then there are concomitantly other quantities (the position of the particle) that can be calculated from these in accordance with causal laws (of classical optics). And the causal principle is *applicable* because these exact measurements can be carried out, since only under the presupposition of these exact measurements can the claim of the causal principle be applied—namely, that there are other quantities connected to the exact quantities through causal laws.

Correspondingly, we find a remark—obviously much less noted—coming from Heisenberg, in which the following is held to be valid for quantum mechanics: "If at any given time certain physical quantities are measured as exactly as possible in principle, then there are also quantities at every other time, the values of which can be exactly calculated, i.e. for which the result of a measurement can be precisely predicted."[5]   38

In this we see that the causal principle is not applicable for all possible events; to the contrary, its applicability is limited by means of the uncertainty relation. And indeed, according to this relation, not *all* quantities of classical physics are fundamentally, and under all conditions, exactly measurable. (Expressed in terms of the formalism of quantum mechanics, this can be stated as follows: The operators corresponding to the observables of the position and momentum of a particle are not commutable. They have different eigenfunctions, and the principal axis systems of the matrices corresponding to the position and momentum operators are not identical.) It follows from this that quantum mechanics does indeed make possible precise statements concerning measurements, as well as precise predictions; but most important, it also contains statements of probability that, on the grounds of the formalism of quantum mechanics, are not reducible to statements that do *not* pertain to probability quantities (*Wahrscheinlichkeitsgrößen*).

Quantum mechanics accordingly falls into two groups of statements: one group for which the above-mentioned causal principle has an application, and another group for which it has no application. Thus, if we formulate the causal principle in the previously given manner, then we find that it is this limitation of its applicability by means of an empirical law, and not the abolition or invalidity of this principle, that differentiates quantum mechanics from classical physics.

The determination of the causal principle by von Weizsäcker also appears to be in agreement with this formulation. He writes: "If some of the determinant elements for the state of a system are known, then we can calculate all of those determinant elements of previous or subsequent states which stand in a univocal interrelation with the known elements in accordance with classical physics."[6] In spite of this, von Weizsäcker does *not* mean—as has just been contended—that the limitation of the applicability of the causal principle in quantum mechanics indicates its difference from classical physics; rather for him the case stands just the opposite. He believes that both would have this limitation in common, since in classical physics as well as quantum mechanics all the determinant elements of a system have never been completely known, owing to errors in measurement and all manner of disturbances. The difference then lies only in the limits to which the determination of a state can be driven.

But in this there is not enough emphasis placed on the fact that the indicated limitation in classical physics is fundamentally different from that in quantum mechanics, owing precisely to the limits mentioned by von Weizsäcker. This is the case because the limits in accuracy of measurement and in information pertaining to measurement in classical physics exist only *practically* (*nur praktisch*)—whereby it is also conceivable that they could be fundamentally and arbitrarily pushed further and further back—whereas in quantum mechanics these limits, in accordance with the uncertainty relation, are viewed as *fundamentally* unsurpassable.

Therefore, whereas in classical physics the causal principle is fundamentally *not limited,* in quantum mechanics, to the contrary, it is fundamentally *only* applicable as *limited.* Hence I believe that we can maintain the difference between classical physics and quantum mechanics stressed earlier.

2.2 The Unlimited Principle of Causality and Hidden Parameters
A determination of the causal principle which contradicts quantum mechanics is, for example, the following: "For *every* event there is a causal explanation."

Since a "causal explanation" is intended here to mean that an event is logically deducible from other events by means of causal laws, for which, however—according to Stegmüller's definition of causal laws—exact values in measurement are requisite, it follows that the statement "every event is causally explainable" means further that every event is exactly measurable. And *this* contradicts the recognized interpretation of phenomena as these pertain to quantum physics.

This causal principle expresses an unlimited claim: that this should be the case for *every* event. Accordingly, I would like to call this the *unlimited principle of causality.* In contrast, that principle which limits itself to only exactly measurable events (and not all events need be exactly measurable, as the uncertainty relation teaches us) I name the *limited principle of causality.*

If one holds fast to the unlimited principle of causality, then from the standpoint of quantum mechanics this means that one assumes exact values *in themselves* (*an sich*)[b] which lie beyond the measurement values and which, in light of the uncertainty relation, are inexact or impossible to determine at all. Under such a presupposition it is then perhaps hoped that these values will eventually be able to be measured or interpolated in some manner or other in the future so as to afford access to the sought-after causal explanation. Today, for the most part, such intrinsic values are called "hidden parameters" (*verborgene Parameter*). But from the point of view of quantum mechanics, can we even entertain a belief in the existence of such hidden parameters and the unlimited principle of causality?

41

Thus, to the two questions initially posed—"How is the causal principle expressed?" and "Is the causal principle valid?"—there belongs unavoidably the third question—"Are there hidden parameters?" Today the camps are still divided as to this question as well.

The so-called Copenhagen school has decided in favor of the rejection of such parameters. The representatives of this school, the most important being Bohr, Heisenberg, and von Weizsäcker (among others), say that it is impermissible to attribute to certain determinant elements of nature an objective sense independent of their respective observational context. The only things given to us are the appearances arising out of measurements carried out in the classical manner and experiments interpreted in the classical sense (classical in the sense previously illustrated in the example of a positional measurement). The complementary elements pertaining to these appearances cannot be connected with the appearances in a world of the in itself. Accordingly, probability statements—like the uncertainty relation—are fundamentally unavoidable. This standpoint, they also assert, will form the foundation for *every* future theory of microphysics.

In opposition to this view, to cite a philosophically enlightening and particularly interesting example, Bohm, in collaboration with Vigier and stimulated by the long dormant yet still fertile thoughts of de Broglie, expounded a theory based upon hidden parameters. Accordingly, this

b. The reflexive phrase *an sich,* placed in italics here by Hübner, has a long history in the German philosophic tradition. The phrase has now become adjectival and adverbial in certain contexts. Here it carries the force of a radical sense of objectivity lying beyond all subjective interpretation or interference. In this form it can be related to almost any substantive: hence, values in themselves, a world in itself, or existing in itself. Unfortunately, used in this adjectival or adverbial form, it is often unwieldy in English. As a result I sometimes translate it as "intrinsic."

42   theory set itself in opposition to traditional quantum mechanics and the so-called Copenhagen interpretation of microphysics.

Bohm[7] initially divided the time-dependent Schrödinger equation (which itself contains a complex function) into its imaginary and real parts, arriving in this way at two equations. One can be conceived as a continuity equation (*Kontinuitätsgleichung*) corresponding to the classical equation of mass continuity (*Massen-Kontinuitätsgleichung*), except that now it states that the probability of finding a particle at a determined location does not change. The other equation, however, agrees with the classical Hamilton-Jacobian differential equation under a certain assumption, namely that Planck's quantum of action h equals zero. If h does not equal zero, then this agreement with the classical equation can be preserved, provided we introduce a new concept of a particular potential added to the classical one.

In this way the Schrödinger equation is interpreted by Bohm as the law for preserving the probability of finding a particle in a particular location; simultaneously, according to this view, it also shows that the dynamic relations of particle movement are described, as in classical mechanics, by the Hamilton-Jacobian differential equation. In this way the paths of

43   particles prove to be classically determined quantities; the wave function represents for Bohm a real field that exerts a force on the particle; thus, according to this interpretation, the occurrences which appeared discontinuous in quantum theory are in fact continuous.[8]

A decision between Bohm's theory and traditional quantum mechanics based on experimentation appeared to be difficult so long as the Schrödinger equation remained fundamental in both systems and the same predictions were forthcoming. For this reason Heisenberg wrote: "Bohm has been able to carry out this idea in such a way that the results for any experiment are the same as in the Copenhagen interpretation. The first consequence of this is that Bohm's interpretation cannot be refuted by experiment."[9]

2.3 The Philosophy of the Copenhagen School and the Philosophy of Bohm

In a certain respect, then, both of the above interpretations could be viewed as equally valid, apart from the fact that they *both* had to struggle to a certain extent with particular problems that have still not been overcome (concerning which we cannot elaborate any further here). But since

44   the resolution of these problems still seemed possible by means of a corresponding improvement in each system's respective formalism, the debate between the different conceptions shifted in part to the realm of philosophy.

Accordingly, both interpretations have also been philosophically grounded and interpreted, and, as might be expected, the philosophy of the one is in clear contradiction to that of the other. I will set their fundamental theses in opposition, beginning with the Copenhagen school.

Bohr and his supporters see a primordial phenomenon of Being in the uncertainty relation: An objective existence could only contain what would be measurable; however, it could not contain what would be complementary to this. Von Weizsäcker holds that there is an ontology at the basis of classical physics which can no longer be preserved today. In this ontology, nature is thought, in Cartesian fashion, to be something existent in itself. However, in opposition to this view, natural laws are not present in nature without an addition on our part; rather they are laws of the possibility of bringing forth or producing (*hervorbringen*) phenomena by means of experimentation. What is (*Seiendes*) ought only to be that which can appear in this manner.

This Copenhagen philosophy can be summarized in the following statement: Being is only the possible, which, with the aid of a measurement procedure, is produced or brought forward as the real.

By way of contrast, Bohm holds that causal laws are intrinsically operative in nature itself. In this view nature is infinitely complex, built up in an infinite number of layers. And each of these only possesses relative autonomy, since each of these layers manifests the effects of a deeper one, the parameters of which remain for the most part hidden.

Bohm summarizes his philosophy in the following statement: "The essential character of scientific research is, then, that it moves towards the absolute by studying the relative, in its inexhaustible multiplicity and diversity."[10]

Which of these mutually contradictory philosophies should one choose? Or might it be that neither of them is convincing? To answer this question, I must illustrate the philosophical aims and tendencies of each of these positions somewhat more closely and consider them critically. I will begin again with the Copenhagen school.

For this school of thought, what is observable on the basis of the uncertainty relation forms the sole legitimate foundation of a scientific assertion. Here "observable" means "measurable." And only that which appears by means of such a measurement is a reality for the Copenhagen school. Thus, according to their interpretation, they believe that the formalism of quantum mechanics only permits transformations from assertions (statements) concerning observables (and thus measurables) to other such assertions (statements). So by refusing to leave the secure basis of "reality," they supposedly quarantee themselves the upper hand in comparison with any theory that works with speculative concepts like unobservable parameters.

For this reason Heisenberg criticized Bohm as follows: "Bohm considered himself in the position to maintain: 'We must not give up the exact, rational, and objective description of individual systems in the realm of quantum theory.' Nevertheless, this objective description revealed itself as a kind of 'ideological superstructure,' which had little to do with immediate reality.''[11]

46   This is obviously the case within the framework of the Copenhagen interpretation, since, there, only that reality which is given in observation can serve as "the sole legitimate foundation of knowledge"; and hence we are not allowed, according to this view, to ascribe an objective sense, independent of the respective observational context, to different determinant elements of nature. All that is *truly* given are the appearances which come to light by means of measurements and experiments; accordingly the complementary structures to these appearances cannot be connected with these appearances in a world of the in itself.

At first glance what we are in fact involved with here is reminiscent of Berkeley's "esse est percipi," now directed primarily, however, against the existence of unobservable parameters. The decisive difference between this view and that of Berkeley lies, to be sure, in the fact that the "esse est percipi" would seem to have been transformed into "Being is being measured" ("Sein ist gemessen werden"). (In chapter 6, where among other things we will treat this difference more explicitly, it will be shown that the statement "Being is being measured" does not really express the state of affairs when viewed more precisely.)

The following should be noted about this empirical orientation: The alleged confinement of physics within the realm of the observable is an illusion; no physical theory would be possible if we exercised this limitation in a rigorous manner—and this is especially true for quantum mechanics.

I would now like to illustrate this briefly.

If the state function Ψ is taken to be a reality for physics in the sense given by the Copenhagen school, then it must be determinable through measurements. But this presents its own problems, since the theoretically devised manner of calculating the Ψ function, with the aid of a partition

47   of a large number of equal systems and a statistical count, cannot be completely carried out for practical reasons.[12]

Now, this could perhaps be interpreted as having no particular significance, since practical problems might one day admit of a solution. However, if we consider problems of this kind in quantum mechanics more exactly, we must confess that such hope already verges on the limits of speculation.

For indeed within quantum mechanics it is possible, by means of an operator, to convert every regular complex function, with certain asymptotic boundary conditions running to infinity, into a function which satisfies the Schrödinger equation. And since, in virtue of the formalism of quantum mechanics, every physical quantity is presented by means of an operator, quantum mechanics would then only be completely interpretable if every operator also corresponded inversely to such a quantity. The set of quantities definable in this manner would then be infinitely large. Hence there are quantities to which no physical meaning can be assigned, as for example the product of the energy and the square root of the momentum. Of course, even such quantities as this could be made interpretable in physics by means of the construction of a special measurement apparatus. However, this would then have to be done at the least for all possible combinations of fundamental quantities of physics, and indeed for all negative and positive potentials as well. From this it follows that an inconceivably large number of measurement apparatuses would be required if every possible quantity in quantum mechanics were to be measurable in accordance with the postulate of thoroughgoing observability. 48

Therefore the idea that the formalism of quantum mechanics allows only for transformations of statements concerning observables into other such statements is far from being adequately founded. Finally, Wigner has proved (cf. chapter 6) that the greater part of the possible operators in quantum mechanics do not represent measurable quantities.

Just as the philosophy of the Copenhagen school proceeded on the basis of measurable observation, so Bohm, for his part, supported his contention on the validity of the unlimited principle of causality. He believed that all probability statements in physics were fundamentally reducible to statements that were not themselves probability statements. For him all probability statements are merely provisional. Again, in his opinion, nature exists absolutely in itself as an infinitely complex manifold and therefore possesses hidden parameters, which, if they could be adequately known, would reproduce the determination of what has actually occurred. For Bohm, therefore, every event is fundamentally causally explicable.

However, the validity of this unlimited principle of causality cannot be proved in a theoretical manner, just as, on the other hand, it cannot be refuted; and this applies to every possible formulation of the causal principle as well.

But what then is the nature of this debate between antithetical positions? How might the arguments both pro and con be founded? Such a foundation could only be given in one of two ways: empirically or a priori. 49

Empirically viewed, however, a causal principle can be neither confirmed nor falsified. Regardless of how we might formulate the causal principle—and there are naturally many more possibilities of expressing it than those investigated in the present examples—in every case, if it is to be at all adequate, it will always be subsumed under a clear logical form which consists of a combined existential and universal statement of the type "for every . . . event . . . there is . . ." And this is the case since, regardless of its particular form, it is derived from the unconditional, universal formulation "For every event there is a cause." But a universal statement does not admit of proof—for how could we know all cases?—

and an existential statement is not falsifiable—for how could we know whether there might not still exist things which we have not yet proved to be existent?[13]

50   Regarding the attempts to found the causal principle in a theoretically a priori manner as something which is necessary (for instance, transcendentally), at least this much is certain: these proofs have remained highly dubious, and thus have found anything but universal acceptance.

51   All of this might be summarized as follows: The causal principle, however it may be formulated, does not represent a theoretical statement at all; it lays claim to expressing neither an empirical fact nor an a priori necessary constitutive structure, whether of nature or of a cognizant being. It is then neither true nor false; rather it implies only the *demand* to *presuppose* and to *seek after* the existent Y for every X. Thereby the causal principle becomes a *practical postulate* (*praktisches Postulat*); thus it is only justifiable in terms of the end for which it is a means. So we no longer need to ask: "How is the causal principle expressed?" and "Is the causal principle valid?" since how the causal principle is expressed is no longer determined according to what is, but rather according to what one wills (the end that is sought). Further, every causal principle is neither valid nor invalid—there is no empirical or metaphysical high court which is able to pass any judgment here. It has no theoretical content; it asserts nothing about the world (which is why it is often held to be a tautology). It is a methodological postulate. Hence, strictly speaking, the two questions must be reformulated so as to read:

1. Which causal principle, regarded as a universal methodological guiding rule, do I wish to lay down at the foundation of physics?
2. What empirical problems do I have to overcome in using this guiding rule?

2.4 Neither the Limited nor the Unlimited Principle of Causality Contains an "Ontological" Assertion; Both Are Established Precepts A Priori[c]

52   In summary we can now see that both the philosophy of the Copenhagen school and that of Bohm proceed on the basis of false presuppositions.

The Copenhagen school holds the observable and measurable appearance to be the single legitimate foundation of knowledge, and the members of this school believe they can support their interpretation of quantum mechanics on this assertion alone—this is *their* error. Bohm, on the other hand, holds the unlimited principle of causality to be an expression of an essential characteristic of the world in itself—and this is *his* error.

c. Again the reader is reminded that *Festsetzungen* (established precepts) is a term that refers to a notion which Hübner will develop more fully in chapter 4 (cf. pp. 42–47).

But both of these views share the common error of seeing the statements and principles of physics as expressions of essential characteristics of nature or being. In the final analysis, they understand physical theories ontologically and overlook the fact that they are really only constructs or models determined by means of a priori precepts and postulates of the most varied sort.

These "apriorisms" should not be confused with those belonging to a metaphysics or ontology. The a prioris of a metaphysics are viewed as necessary—here Kant's synthetic judgements a priori might serve as an example. The a prioris of physics, on the other hand, are in no way necessary; they can be replaced by others.

The different causal principles discussed here bear witness to this, as do the hidden parameters of Bohm. But there is yet something else which bears witness to this as well, namely, the fact that it is equally possible to have different theories for the same realm of experience. *Truly decisive* arguments of a physical or philosophical nature cannot be brought forward in opposition to either the one or the other position. It appears to be an inextirpable characteristic of people to transform everything which in truth springs from their own invention and design promptly into an objective givenness. The history of physics is a process in which this confusion of our own free constructions with the ontologically real constantly repeats itself.

As soon as the absolute reality of Aristotle's doctrine of proper place was abolished and the doctrine of motion was connected with the free choice of position, we find that a new absolute reality was immediately introduced, namely, that of inertial motion. This motion was not supposed to be dependent upon the choice of the reference system or body; rather it was held to be an essential and true constituent characteristic of the system itself to which it was attributed. This was still conceivable within a Cartesian framework, as chapter 9 will show. But in the later interpretation of the principle of inertia, one had to realize that this presupposes the concept of "equal times," the criterion of which is again the law of inertia; for "equal times" should indeed be present if a body, free from external forces, traverses equal distances. In this way it was revealed that inertia is neither a necessary nor an empirical characteristic of the thing itself, but rather something which corresponded to a freely chosen precept concerning the standard of measurement: the principle of inertia became a definition of measurement (*Maßdefinition*).

The freedom of a priori precepts that comes to light here offers us, in fact, the key to understanding how to eliminate the dogmatic strife among metaphysical formulations built upon particular physical theories or guiding the construction of such theories. It does this by indicating that none of these theories can claim to express *the ontological* structure of the world, because all of these theories are only *possible* interpretations which

53

54

have practical postulates at their bases. But then the *proper problem* posed by physics for philosophy also has to do with this freedom rather than with some questionable model which is always necessarily ephemeral.

What exactly is this freedom? A large part of the following investigation will be concerned with this question. In addition, in chapter 6 the question of the hidden parameters in quantum mechanics will be amply illuminated from other perspectives than those dealt with here. But now I want to detach the question of a priori precepts and foundations in physics from the particulars of quantum mechanics and treat it in a more universal and systematic manner.

3

SYSTEMATIC DEVELOPMENT OF THE
QUESTION OF FOUNDATION IN THE
NATURAL SCIENCES

Belief in facts (*Tatsachengläubigkeit*) is a characteristic of the modern   55
world. This belief—like every other—demands that the believer bow to
that which is believed. In this case it would then mean: "Bow down to
the facts!" The fact is held as something absolute which speaks for itself
with coercive power. Experience can thereby easily be compared with an
appellate court to which an appeal is made and from which a judgment
comes. And like every high court, this too is looked upon as an objective
high court. But that realm commonly thought to possess such objectivity
is science; and hence science is held to be the guardian and discoverer
of truth.

What is there in these opinions which might be accounted correct?
What is the actual status of the foundation of science on the basis of facts?
Let us again begin with an example of what passes today as the ideal
model for most of the sciences: a physical theory.

Such a theory consists of a group of axioms which take the form of
differential equations with derivations of state functions at a world point
with respect to the time parameter. Natural laws are deduced from the
axioms and then brought into a unified interconnection within the framework of the theory, where they are also ordered among one another and
classified. By setting boundary conditions (*Randbedingungen*), inserting
measurement quantities for variables, we obtain the so-called basic statements (*Basissätze*) that correspond to this theory. From these, other basic
statements are then deduced by way of theorems pertaining to the theory,   56
which predict what the measurement results at a later time will be—and
these can then also be checked by measurements.

It thus becomes clear that these basic statements can be seen as the
empirical foundation of the theory—for which reason they are called "basic
statements": It is *these* which should express the facts that the theory
must call upon for support; the sought-after objective verdict must be
expressed *in them; they* must occasion the requisite empirical decision as
to whether the theory is true or false, whether or not it corresponds with
nature.

25

Therefore let us first examine precisely to what extent basic statements express the facts, and then to what extent these facts can found natural laws on the one hand and the axioms of the theory on the other.

3.1 The Foundation of Basic Statements

A basic statement expresses either an obtained or an expected measurement result. Instruments are required for this measurement. However, in order to use such instruments, in order to place trust in them, we must already have a theory concerning the form and manner of their functioning. This holds even for the simplest instruments—as for example a ruler (any standard of measurement) or a telescope, since when we use a ruler we obviously presuppose that by changing location it undergoes no change or a calculable change (thus presupposing a certain metrics), and if we use a telescope this requires the acceptance of certain ideas such as how light rays behave in particular media, etc. (i.e. it presupposes a certain optics).[1] But if the measurement procedure is to have the meaning (*Sinn*) given above, then both a theory of the instruments employed and, *moreover,* a theory concerning the quantities to be measured must precede the measurement process, since the concept of the quantities in question is not given through some indeterminate kind of everyday experience, but rather can only be defined and determined in terms of a theory.[2] A brief example: If we wish to carry out a measurement of the length of light waves, then we must in advance have a wave theory of light; further, with the aid of this theory and a theory concerning the measurement instrument, we must know how this instrument makes possible such a measurement of wavelengths; and third, we must know how to read the sought-after measurement value from the instrument.

Thus we see that the basic statements which should express the facts upon which a theory is founded in no way communicate pure perceptions (such as gauge readings, congruences, displacements, etc.); rather, these basic statements also have a theoretical content. Basic statements do not say: "I have this and that perception," but rather: "This and that wavelength, amperage, temperature, this and that pressure, was measured, etc." And all these concepts have meaning and content only within a theory.

Finally, because the exactness of measurement always exists only within limits, every measurement allows for the arbitrary reading of various measurement data within certain limits. When we choose one of these, it is not a matter of experience or perception, but rather of decision. As a rule this does not occur capriciously, but instead within the framework of a theory of error analysis. But such a procedure of error analysis does not alter the state of affairs, since, among other things, the following nonempirical presuppositions underlie such a theory: the acceptance of the existence of a true mean value; the acceptance that the errors have with equal probability both a positive and negative sign. Further, one must assume as a precept (*Festsetzung*) that the analysis of error is determined with respect to the squared deviations from the mean value, etc.[3]

Thus the following is shown: A basic statement does not express a pure fact and is never necessitated by such a pure fact; it cannot be an extra-theoretical foundation for a theory; it is itself theoretical, determined by means of interpretations, and something which arises out of certain decisions. 59

3.2 The Foundation of Natural Laws

To what extent, then, can basic statements found natural laws? Let us ignore for the moment the recognition that basic statements do not express absolute facts, and let us assume that, despite this, they are looked upon as if they were adequately empirically determined—which is in fact the way they are generally viewed. If we assume this view, then the foundation of a natural law must be sought in the following manner: Measurements are made; from these a curve is produced that presents the mathematical function in which the natural law in question is expressed; the curve, so it is asserted, substantiates the law. But the production of such a curve can never occur on the basis of measurements alone. Since measurements can only yield sporadic results, it is always necessary to call on interpolations or smoothed-out data (*Glättungen*), which for their part again only rest upon decisions and precepts. Accordingly, we have here a state of affairs analogous to that found in the calculation of error. Without these precepts there is no foundation for natural laws on the basis of measurements; and with them, this foundation is once again something which can claim no necessary basis in facts.[4] 60

But let us now consider the interrelation of basic statements and natural laws more precisely. Constants of nature play a decisive role in natural laws. And even admitting the role which might be played here by smoothed-out data, interpolations, and theoretical presuppositions or decisions in the determination of such constants, many people nonetheless point to the fact that there is a relative equality in measurement results, even when these results are obtained in the most varied of ways. Regardless of the manner in which we might approach such things, they always yield a numerical equality. Therefore all of the silently made presuppositions detailed above appear in retrospect to have been justified and transformed into facts.

Before we examine this general assertion, let us first give an example which better illustrates the point. The example concerns two different methods for determining the speed of light: in the first case, by means of the constant of aberration (*Aberrationskonstant*); in the second, by means of Fizeau's method. Both of these lead to the same result, though they rest on completely different measurement procedures. Let us then first examine the manner in which nonempirical presuppositions enter into both of these measurements.

61   If the constant of aberration is known, then the speed of light can be calculated, provided the velocity of the Earth is also known. However, the velocity of the Earth can only be determined for its part if the distance it travels in a particular time interval is known. Thus two time measurements enter into the calculation of the speed of light: one at the beginning of the time interval and one at its end; and both of these time measurements occur at different locations. But this means that we must presuppose that the clocks used here are synchronous and that their movement remains constant. In order to measure the velocity of the Earth, we must therefore determine when two events, separated by distance, are simultaneous. But at least since the time of the appearance of the relativity theory, it has been recognized that the simultaneity of events separated by distance is not an observable fact, and thus that such a determination rests upon precepts. In this way we then come to see exactly which precepts are involved in the method for measuring the speed of light with the aid of constants of aberration.

Now let us consider Fizeau's method for measuring the speed of light. He projects a beam of light over a known distance to a mirror, from which the beam is then reflected back to its point of origin. He can then calculate the speed of light by determining the time interval between the initial projection and the arrival of the returning reflection. Here the precept involved has to do with the fact that the speed of light is taken to be the same in both the trip to the mirror and its return. If we wish to transform this precept into an empirical fact, then we would supposedly have to measure the time elapsed from the light beam's emission to its reflection, as well as that between the emission and the absorption of the returning

62   light beam. But then we would again have two time measurements of events separated from one another by distance; accordingly the precept would enter in at another place.

This pointed example should have provided us with the answer to the following questions: Do precepts, which enter into measurements, determinations of constants, and the foundations of natural laws in a fundamental manner, retroactively show themselves to be facts because they lead to the same results, though the precepts used are independent of one another? And can we thus infer retroactively from the agreement in results the empirical truth of the presuppositions which make these results possible? When more precisely stated, such inferences would have the following form: The precepts, independent of one another, P_1, P_2, . . ., P_n, constantly yield the same net result R; *therefore* P_1, P_2, . . ., P_n are empirically true. However, nothing justifies this conclusion. For since the net result R is itself in no way immediately given, occurring in each particular case only on the basis of precepts, the *only* thing we can assert is that the indicated agreement is also merely the result of precepts. Thus all we are entitled to say is that precepts, which lead to conformities of agreement in the ways described, are purposely chosen because they lead to a certain simplicity in physics—and nothing more. The difficulty in recognizing this simple state of affairs is grounded solely in the fact that we constantly carry around within us a metaphysics according to which statements of physics must describe, in some way or other, an intrinsically existent reality.

From all this it then follows that neither basic statements nor natural laws express straightforward facts in any sense at all; rather they are codetermined by spontaneous decisions.

3.3 The Foundation of Axioms Pertaining to Theories of Natural Science

It would seem to follow from the above that it is unnecessary to ask about the empirical foundation of the third group of propositions belonging to a theory, the axioms. However, as before, in the examination of natural laws, we will again ignore the previously developed results and assume for the moment that they are not valid. Let us now concentrate only on the logical fact as such, namely, that the axioms, the very heart of a theory, are the premises from which the basic statements are deduced as consequences. It then follows that if the consequence is true (here, if the basic statement predicted by the theory is confirmed by a measurement), then, in accordance with the rules of logic, the truth value of the premise (in this case, the axiomatic system of the theory) is not established. It can be true, but it can also be false. From this, moreover, it obviously follows that many various axiomatic systems can be devised for the same basic statements, even if these basic statements can be interpreted differently in terms of different theories. And with this the question is raised—analogous to the question previously discussed concerning different methods leading to the same results—as to whether it is possible to construct something like empirical facts on the basis of the competition between such different theories. Thus far we have only examined the empirical possibility of foundation (*empirische Begründbarkeit*) for any single theory whatsoever; now we turn to the treatment of groups of theories. And here we consider the following possibilities for the comparison of theories. (This will be treated further in chapters 5, 6, 11, and 12.)

1) The theories have the same basic statements B—which, however, may be interpreted differently within their respective frameworks—but one of the theories is the simplest or deals with further statements B'.
2) The theories are structurally equal.
3) One of the theories contains the other as its special or limited case (*Grenzfall*).

All three of these possibilities have been used in order to give the criteria for the factual content of theories. Let us begin with the first.

Here it has been asserted that the simplest or most comprehensive theory is the true one or the one which approximates the truth most nearly.[5] This view presupposes that nature *is* to be construed simply or comprehensively (not to mention the presupposition that nature is to be construed in terms of that "simpler" or "more comprehensive" theory being dealt with at the time!). But how can this be proved if, as stated, the theory in question, which would indeed confirm and reveal this kind of constitution for nature, cannot be proved true in its own right?

Concerning the second possibility, it has been asserted that if several theories have to do with the same basic realm (*Basisbereich*), then they must have the same structure—and *this* is then taken to be the empirical truth.[6] But what exactly does structural equality mean? Briefly stated, two sets have exactly the same structure when the following conditions are fulfilled:

65 1) Each particular element of one set can be coordinated in a univocal manner with each particular element of the other set.
2) If certain elements of one set are related in a particular manner, then the coordinate elements in the other set are also so related.

From this it follows that if two sets—both of which consist of a system of statements, as in a theory—have exactly the same structure, then statements of one theory can be deduced from statements of the other theory and vice versa. But this is precisely what is *not* necessarily given in the case of two theories that have been devised with respect to the same basic realm. The only thing which these theories have in common is the basic realm itself; but this says nothing about their structural equality. Since such a structural equality does not as a rule occur in the given case of competing theories, the structure of theories cannot then present us with an invariable empirical factual foundation.

Concerning the third possibility, it is asserted that as a rule theories ultimately become the special or limited cases of other theories, and further that scientific progress consists essentially in this. This occurrence is also taken as proof that theories are built up out of facts, since that theory which becomes the limited case might indeed be enlarged upon and brought into a greater systematic interrelation and completion by the theory which contains it, while taken in itself the original theory is irrefutable because it rests on facts. Here it will be sufficient to examine the relation of Newtonian physics to the special theory of relativity, since it is this relation which is generally held to be the classic example for the above-mentioned assertion.

Even today many physicists maintain that Newtonian physics is a lim-
66 ited case of the relativity theory and that as such it has to do with that

province in which velocities are substantially less than the speed of light. This view is grounded in the assumption that this limited case can be deduced from the theory of relativity.

But what would such a deduction look like? If we indicate the statements contained in the special theory of relativity with $R_1 \ldots R_n$, then, in order to derive Newtonian physics as the limited case of this, we must add a statement of the following kind: In Newtonian physics $(v/c)^2$ is significantly less than 1. In this way we would then arrive at statements $L_1 \ldots L_n$ (L stands for a value significantly less than 1); it is only in this sense that we could speak of a derivation of the one theory from the other. Thus viewed, these L_i are indeed special cases of the special theory of relativity; however, they do not comprise Newtonian physics, nor are they special cases of this, since the variables and parameters representing the location, time, mass, etc., in the R_i system admit of no change in the L_i. They are different from the classical quantities, though these have the same names. The concept of mass in Newtonian physics refers to a constant; that of Einsteinian physics is interchangeable with energy, and thus variable. Newtonian physics defines space and time as absolute quantities; the opposite is the case with Einstein, etc. This clear logical difference allows for no deduction of the one theory from the other, though again the same terms are used in both. If we accept no transformed definitions, then the variables and parameters of the L_i do not belong to classical physics; and if we do redefine them, then we can no longer speak of a deduction of the L_i from the R_i. In the transition from Einstein's theory to that of classical physics, not only does the form of the laws change, but the concepts upon which these rest have been altered as well. Therefore Newtonian physics is not a limited or special case of Einsteinian physics. The entire revolutionary force of Einstein resides in his new definitions.[7]

The general theory of relativity and the Newtonian gravitational theory also show the same logical incompatibility with one another. According to Einstein, the universe is curved and without gravitational force; Newton's universe, on the other hand, is a Euclidean space within which gravitational forces act. And apart from the reasons just given for why it is inappropriate to hold that the general theory of relativity might be taken to contain Newtonian physics as a limited case (for example, in cases of relatively small and therefore insignificantly curved areas of space), we must also take into account that the Newtonian theory—with few exceptions—extensively describes and predicts astronomical movements just as correctly as the Einsteinian theory; and this holds not only in those limited cases mentioned above, but in all other cases as well. Hence there can be no talk at all of the Newtonian gravitational theory as the limited case of the general theory of relativity which has superceded it.

From this it follows that in connection with the mutual relation of competing theories, one theory *need not* contain the other as its limited case; nor is this even generally the rule. We cannot even correctly assert that one theory approximates the other, for in this the *tertium comparationis* is lacking. But then how is it even possible to say that the measurement results in certain cases turned out to be similar or equal—since this would indeed indicate such an approximation—if the quantities read from these measurements have a logically different sense (*Sinn*), in the manner indicated above?

3.4 Only Metatheoretical Statements Can Be Purely Empirical

This purely logical analysis of a physical theory and its relation to other theories (which will be examined more deeply in the following chapter) deprives all attempts at giving necessary criteria for the empirical verification of a theory of their firm footing. Such a verification is shattered in advance by the fact that a theory contains universal statements, though not all particular cases are provable. And it then veritably loses all meaning when we recognize the role definitional precepts play in such verifications, when we come to see the merely indirect connection which exists between verification and observation and perception within the measurement process itself, and finally when we consider the fact that mutually contradictory theories can replace one another reciprocally.

But what, then, is the status of an empirical falsification of a theory? Until now we have only considered the possibility of founding a theory, its confirmability (*Bestätigungsfähigkeit*), by means of facts. Might it not at least be possible to see exactly when the theory does *not* correspond to the facts? But since these facts, which are to serve as a strict judge, do not exist at all—as should now have been shown—they can indeed neither substantiate nor confute. Thus both the acceptance and rejection of a theory rest just as obviously on nonempirical decisions. Nevertheless, let us now consider the falsification procedure more closely.

If we ignore the case of internal contradictions within a theory, then this falsification procedure can only consist in the fact that one or more measurement results contradict at least one prediction that is deducible from the theory. As a rule the theory is seen here in such a way that the exactness anticipated in measurement, the probable limits of interpolations of measurement results and of interference, are all calculated into it in advance. This means that we cannot attribute deviations in the anticipated results to inexactness of measurement or to inappropriate interpolation or external interference which is not interpretable by means of the theory. Thus these deviations are to be viewed as falsifications of the theory. But is this *empirically* necessary? Are *empirical* facts the compelling force behind falsification?

Even if someone *decides* not to give up the theory despite predictions that have failed, if he says that there is interference which is not dealt with or interpretable in terms of the theory, and that such interference is responsible for the negative result, that there are ad hoc supplementary propositions which save the theory, errors in the measurement procedure, etc., it will still have to be admitted that all these "there are" statements are not empirically falsifiable as such, and therefore not *empirically* refutable. If these are to be refuted, then this can only occur by indicating that it would be *methodologically* inappropriate and methodologically unreasonable to base one's hopes upon them. Thus, when Popper, for instance, asserts that it is better to seek the falsification of a theory than to attempt to salvage it, he does so with an eye to a good method and not on the basis of some kind of absolute facts.[8] Here I wish to designate such general methodological suggestions as methodological postulates. But are Popper's methodological postulates *always* really appropriate and reasonable? Chapters 5 and 10, in particular, will show that this is not the case. 70

If, however, there is neither empirical verification nor empirical falsification in any strict sense, then the question arises as to whether empirical facts play any role at all in the construction, acceptance, and rejection of physical theories. The answer is that they do. But we need to identify the place empirical facts occupy in terms of the considerations represented here.

With the aid of nonempirical precepts P, we obtain measurement results M, expressed in terms of basic statements. But by using other such precepts P', we get other measurement results M'; and *this* occurrence—namely, that with one set of precepts we get one set of results, while with another set of precepts we get other results—*this* is an empirical fact. If we now add in other precepts, we get statements that express natural laws N; and again with the addition of a different group of such precepts, we get natural laws N'. *This* too is an empirical fact. Moreover, the theory T built up in this way is again a matter of mere precepts. Now, if we proceed on the basis of this theory and make measurements within the context of the theory, then we might find that the precepts P yield the measurement results M, which, on grounds of the previously indicated methodological postulates, would force us to declare the theory falsified; on the other hand, with other precepts P', we arrive at results M', which, again on grounds of the same postulates, would not force us to a falsification. If we proceed on the basis of another theory T_1, then the same thing will happen again. But whereas before we arrived at measurement results M, M', by using theory T_1 instead, we will get results M_1, M_1'—and *this* is also an empirical fact. 71

From this it follows that the *contents* of theoretical statements are not empirical: neither P nor N nor T nor the basic statements pertaining to measurement results M manifest themselves as empirical facts. In this way the only purely empirical element is seen to be the *metatheoretical* inference structure: "If such and such precepts, postulates, theories (all of which are metatheoretical denotations)—then such and such basic statements, falsifications, or verifications (and these are also metatheoretical expressions).'' Or formulated another way: "If we have such and such statements—which say nothing about nature itself—then such and such other statements follow empirically—which again say nothing about nature.'' Empirical facts are indicated only in *these* metatheoretical if-then relations; but the content of the statements pertaining to the theory itself does not present an empirical state of affairs in any sense at all: *Reality first appears not in the theory, but rather in the metatheory.*[9]

Thus far we have only demonstrated in a general manner that different a priori precepts necessarily belong to a single empirical theory. In the next chapter these precepts will be organized and arranged according to categories. However, there we will also broach the question—and give an initial provisional answer—as to how these a priori precepts might properly be justified; in other words, we will ask whether the freedom associated with these a priori precepts points us toward a more profound insight or merely indicates a state of capriciousness.

72

4

AN EMENDATION OF DUHEM'S HISTORISTIC THEORY OF SCIENTIFIC FOUNDATION

It is noteworthy that the theory of science coming into prominence around   73
the turn of the century was still closely tied to the study of the history
of science. Names like Mach, Poincaré, LeRoy, and especially Duhem
clearly bear witness to this. However, this development ceased to follow
the path opened up for it by these men. The historians separated themselves from the philosophers and, in general, received little attention. The
ruling conviction was that the task of the historians was purely "museum
work."

This view seems to have arisen principally from the fact that the object
of natural science which was the major topic of study for the theory of
science at the time—namely, nature—was viewed as a nonhistorical entity
that could be progressively investigated in an increasingly exact manner.
To this end it was thought that it was only necessary to devise the corresponding methods for the formulation, justification, examination, and
application of theories. These methods were then treated as a function of
the unchanging entity to which they referred and were thus themselves
considered to be essentially unchanging in the same manner, even though
they admitted of constant improvement. The theory of science, it was
believed, also advanced continually, just as natural science itself. This
theory was thought to be the product of an abstract acumen for which
direct analysis of certain important theories of the time, such as quantum
mechanics or the theory of relativity, sufficed. The historical perspective,
especially as this related to fields outside the province of physics, was   74
thought to yield very little. Indeed, it was expressly protested that it is
of little interest what scientists *have* done; rather it is much more important
to ascertain what they *should* do. And this was supposed to express the
idea that the theory of science needed to create a universal organon of
science in general, something like formal logic. Today this is still a widely
held opinion.

In opposition to this, I propose here the thesis that the study of history
is of decisive importance for the theory of scientific foundation. Indeed,
such a theory is not possible without historical thought.

35

4.1 Duhem's Historistic Theory of Science

The first person to embrace this point of view was Duhem.[1] Duhem declared that he could be a physicist only as a theorist of science and a theorist of science only as a historian of science. Experience in both the teaching of science and scientific research had made this insoluble interdependent relation clear to him. The impossibility of building up a physical theory piece by piece with logical and empirical necessity, the obscurities and the confusion which accompany such an attempt, forced him to reflect about the theory of such a theory. His conclusion was that the justification 75 of a system of physics could only lie in its history.[2] For this reason I would like to call his theory of science "historistic." I will begin with a brief presentation and interpretation of his philosophy.

The point of origin is the insight that only by means of a complex transposition mechanism (*Uebersetzungsmechanismus*) are we led from the givens to the statements or assertions of a physical theory—a mechanism which does not allow for a univocal coordination of these two. This point was expressed in the previous chapter, though there we were using means more modern than those at Duhem's disposal. The meaning of this can be summarized as follows: It is possible that countless mutually exclusive theoretical statements can correspond to one and the same fact because of the limitations inherent in the accuracy of measurement (such as the reading of instruments).[3] Also, when dealing with quantities to be measured, we are concerned with concepts that do not rest upon abstractions drawn from what is directly perceived: Concepts like "tree," "sun," and "river" are quite different from those like "electron" and "electromagnetic wave" because the latter are only understandable within the framework of complex physical theories and are mediated by the context of these theories.[4] Further, we can only understand the function of instruments used for such measurements and justify their reliability by presupposing the very theories upon which they rest and according to which they were constructed.[5] For these reasons an experiment concerning an isolated hypothesis can never be decisive, because whatever the 76 result of an experiment may be, it will be dependent upon an entire system of theoretical assumptions, assumptions which can in no way be adequately examined in isolation from one another.[6]

Therefore Duhem concludes that, while a theory can indeed fail, whether this is actually the case will depend upon the criteria used in the selection of the transposition mechanism that leads from the givens to theoretical assertions. And these selection criteria, although indispensable, are not given with necessity either by nature or by a universal form of reason. Thus here once again we touch upon that freedom already discussed in chapters 2 and 3.

The more this insight forced itself upon Duhem by way of the analyses he had undertaken in terms of both his work in scientific research and his teaching, the less it was capable of satisfying him. In no way did this answer the *quaestio juris*[a] of a physical theory; to the contrary, this *quaestio juris* was presented anew and all the more acutely. Did physics dissolve into pure caprice when the assigned selection criteria were not established with necessity? Was there then no objectively binding high court for deciding about the acceptance or rejection of physical theories?

As I have previously stated, Duhem found this high court in the history of science. This alone, according to him, allows a physical theory to appear intelligible at all, and thus grants it the possibility of its entire analysis. Only a theory of science that proceeds abstractly, nonhistorically, and therefore incompletely gives the impression of unlimited freedom in the choice of transposition mechanisms. In opposition to this, according to Duhem, the history of science allows us to follow what are almost always the well-founded steps of the development which has led to the formulation and acceptance of theories. Indeed, though none of these steps was occasioned by any form of necessity, nonetheless the presence of a certain *bon sens*[b] belonging to physics seemed to him unmistakable in history.[7]

Duhem understood this *bon sens* as partly historical and partly nonhistorical. It was historical insofar as it remained tied to a particular historical situation, that is, insofar as we must study all the details and ramifications of a situation in order to be capable of intelligibly appropriating its original insight. But no universally binding rules separable from the situation can be obtained in this way. Sometimes (e.g., despite experimental difficulties) this *bon sens* will retain foundations for a theory which are not directly examinable or provable, while at other times it will give up those foundations, which had been uncritically carried along, and replace them with new ones. This *bon sens* does not seem to concern itself with universal rules of falsification or verification like those put forward so readily today. And decisions of this and similar kinds are always justified only through the particular, historically unique context in which the theorist finds himself.

Thus "historical" does not mean that something once held to be true later proved itself false. This is not the intended meaning. *Here, rather, "historical" means that the picture of nature drawn by physics, insofar*

---

a. *Quaestio juris* is a question or problem of right or justification. The concept has a long history in philosophy and is directly related to the original meaning of critique. In Greek the *arche kritike* is the power (or body) of decision making or judgment. It is also interesting to note the relation to Kant's *Critique of Pure Reason,* in that Kant develops the notion of critique in direct reference to a judicial proceeding, that is, a "tribunal" (cf. Kant, *Critique of Pure Reason,* A xi–xii). The very title of Hübner's work points to this relation as well. Whereas Kant deals primarily with the *quaestio juris* of the metaphysical employment of reason, Hübner is more directly concerned with the *quaestio juris* of the scientific employment of reason.

b. The French expression *bon sens,* used by Duhem, has the meaning of "common sense" or "good sense."

as and to the extent that it is only the function of a transposition mech-anism, is one which arises out of a specific situation and disappears again with this situation. From this perspective the picture of nature is itself only an integral constitutive element of history—which is to say that it does *not* refer to an eternal prototype, however this might be understood, which it then more or less approximates. We must understand Duhem's intention in this way, even though he did not express it in these words.

78

For Duhem it is a classical error to lose sight of the historical condi-tioning of established theoretical judgments and then to hold these to be universal, eternal, self-evident truths. He develops this position through the use of a number of examples, among them that of Euler.[8] Euler be-lieved that the principle of inertia rested upon an insight of pure reason and that this principle impressed itself directly upon the mind of the unlearned person. Thereby he overlooked the fact that such apparent evidentness was principally the result of a protracted historical process, each step of which was contingent, the result, as it were, of a gradually instituted habituation to rules created through tedious piecework and end-less discussions. Aristotelianism, which ran counter to this principle, would have been more readily able to support itself on the basis of such an unmediated insight (although this also would have only applied condi-tionally).

To what extent, however, is this *bon sens* of physics also nonhistorical? For one thing, according to Duhem, it is always led on by the same feeling and belief—which means for him that the classifications which follow from a theory reflect an ontological order.[9] (Thus, for example, under such an arrangement the appearances of light refraction could be assigned to one realm, whereas the bending of light could be assigned to another.) Obviously, for reasons already stated, the true and unmediated image of this ontological order could not be given in such classifications; but for Duhem one nevertheless had to *believe* in this analogical relation, if every endeavor of physics were not to be condemned to the status of a mere play of shadows.[10]

79

Further, this constitutive belief in an ontological order leads to some invariable guiding rules for the theory of science which, as Duhem be-lieved, run through the entire history of science like a red line. These guiding rules call for the development of an ever greater unity and uni-versality in physics. And only because such guiding rules have supposedly been followed does the history of physics present itself as a chain of continuous evolutions. In this way physics supposedly moves gradually, piece by piece, toward the construction of an increasingly comprehensive whole. Duhem believed that it was principally with an eye to this goal that the individual physicist reworked the body of physics he found pre-sented to him, and thus marched forward in his research. Therefore, for

CHAPTER 4

Duhem, the distant, and perhaps even unreachable, ideal was that of a theory with a few axioms from which all known, as well as yet unknown, appearances could be deduced.

Duhem attempted to demonstrate the nonhistorical intentions of this *bon sens* through the example of the history of Newton's gravitational theory.[11] In his view there was a continuous development, guided by such constant intentions, leading up to the formulation of this theory. He presents this development as follows:

Aristotelianism conceives of a point at the center of the universe as the *oikeios topos*[c] of heavy bodies. For Copernicus, in opposition to this, there is a universal striving of all parts of all bodies—thus also for heavenly bodies—to remain with one another and to order themselves spherically. Gilbert, foreshadowing a still greater unification, sees a model for this striving in the magnet. Kepler and Mersenne move even further in the direction of universalization and postulate the gravitation not only of parts of heavenly bodies to one another, but also of one heavenly body to another. This thesis was corroborated on the basis of the observed ebb and flow of tides. Roberval speaks at the same time of a universal, all-encompassing reciprocal attraction. However, Kepler, Bullialdus, and Kirchner, almost simultaneously, had already recognized that this attraction had to be a function of distance for simple reasons. Borelli, in this respect a precursor of Huygens, reasserts the ancient view that a centrifugal force kept the universe from collapsing into a *single* star. Hooke recognizes, in a furthering of Kepler's thoughts, that the gravitational force must stand in an inverse relation to the square of the distance. And Newton finally solved only those unresolved mathematical problems that still stood in the way of the synthesis of all these hypotheses into a unified theory.

Thus everywhere Duhem looked he saw a constantly developing unity, hence an evolution, continuity, universalization, organizing classification—in short, *bon sens*. In this sense there exists an accord between Duhem and the theories still in use today.

4.2 Critique of Duhem's Theory

However, Duhem's presentation of the history of the gravitational theory rests on marked simplification and gives a distorted picture. His chosen example can be turned against him and used to demonstrate that *bon sens,* as he understood it, becomes meaningless, and therewith that Duhem's thesis about the historicality of scientific-theoretical principles, precepts, and physical theories is actually applicable in a much broader sense than he had supposed.

c. *Oikeios topos* is the Aristotelian doctrine of "proper place."

When Copernicus concretizes the Aristotelian *oikeios topos*,[d] this does not correspond, as Duhem supposed, to the intention of arriving at an evolutionary advance through a greater universalization, but is rather, in opposition to this, the result of a previously thought-out decision on Copernicus's part to stand the time-honored and then still accepted physics and cosmology of his day on its head. And it is precisely in this that the famous Copernican revolution (*Kopernikanische Wende*) consists. This radical revolution cannot be explained on the basis of genuine physical or astronomical problems alone; consequently it is not at all principally the work of a *bon sens* belonging purely to physics. As extensive research has in the meantime shown, it is rather the case that the entire upheaval of the Renaissance played a part in the decision as well.[12]

One might be tempted to relate the idea created by Copernicus in the spirit of humanism—namely, that the universe must be construed according to principles of simplicity—with Duhem's notion of *bon sens;* but it is quickly proved that this attempt could not succeed. First, the unity that Copernicus construed did not develop out of the previous system in a continuously evolutionary manner. Second, this example shows pointedly that Duhem's ideal of unity in no way possesses a nonhistorical content as he supposed, because the lack of unity that Copernicus so sorely contested in Aristotelianism, with its all-encompassing division between the higher and the lower, the heavenly and earthly, was not generally considered disturbing; to the contrary, it was viewed at the time as the expression of a divine ordering of things. For this very reason, Ptolemy had also rejected the consideration of a greater unification on the grounds that such was merely formal. The fact that Aristotelianism could have been accepted for so long in the face of the difficulties it offered can only mean that the "revolution" which appears here must be seen as a historical event and not as the result of some insight into certain essential structures belonging to human reason.[13] Third, the idea of unity, which floated alluringly before Duhem's gaze, can scarcely be the topic of discussion in the case of Copernicus; for he unscrupulously purchased a greater unification in astronomy at the expense of a greater disunity in physics. Indeed, Copernicus brought his strictly physical arguments into play only as ad hoc support for his new world-system. These arguments represent the mere reversal of corresponding Aristotelian arguments, without, however, being metaphysically grounded as were the latter. Thus, when we investigate the matter more closely, little remains of the Copernican idea of unity which had been so quick to spread. Indeed, instead of arriving at thirty-four epicycles as his theory asserted, in truth he needed as many as forty-eight.

d. This "concretization" (*Materialisierung*) refers to Copernicus's use of the sun as a material instantiation of the *oikeios topos*.

What then immediately followed Copernicus was almost the opposite of what could be called *bon sens,* and this most precisely in those areas where the most fruitful advances in science were made. When the conflict surrounding the Copernican system is studied in detail and seen within its immediate historical context, it must be said that the rapid propagation of this system mocks Duhem's notion of *bon sens* in physics much more than it pays homage to it. Further, this propagation took place for completely different reasons than those indicated by Duhem, reasons lying outside the realm of physics proper. (This will be the subject of a more detailed discussion in chapter 5.)

In this regard, it is significant to note that the breakthrough to a completely new conception of space, that conception first capable of giving a suitable foundation to the Copernican system, came first from Giordano Bruno, and thus from a philosopher. Here I refer to the breakthrough to the concept of an infinite, homogeneous, isotropic universe. The grounds upon which Bruno based his view were of a purely philosophical nature. The same thing can be said for Descartes, who first brought Bruno's idea to a conclusive victory, since he understood the identification of Euclidean space with that of physics, and hence with matter, to be a postulate of self-certain reason. Moreover, it was upon this decision, and this decision alone, that Descartes formulated his entire physics. Hence it was Descartes's decision in favor of rationalism and for the employment of rationalism—and not merely the further thinking through of genuine problems of physics—that brought about the new Cartesian revolution in physics. When Newton finally succeeded in supplying a provisional conclusion to the numerous attempts at saving the Copernican system—attempts which had been made over and over again on the basis of entirely new presuppositions and grounds—he did so only because he too proceeded on the basis of a metaphysical idea inherited via More and Barrow, namely, the idea of absolute space as distinct from matter and absolute time as distinct from movement.

Therefore, when the steps which, according to Duhem, led to the formulation of the Newtonian gravitational theory are viewed in their true connections, we arrive at a completely different picture of matters than that given by Duhem. None of the new approaches that I have mentioned in this synoptic glance back into history come into being evolutionarily or continuously from that which preceded them by means of a piecemeal assembly into ever greater unities. On the contrary, I believe that the expression "scientific revolution" is fully appropriate to each one of these. And in the above we also see that the causes for such new approaches are often not at all of a genuine physical-theoretical nature (*physikalischer Natur*), but rather can only be deduced from the entire cultural context (*gesamten geistigen Lage*) within which they occurred. (We will return to this point in chapter 8.) Nevertheless we are still able to speak of

AN EMENDATION OF DUHEM'S HISTORICIST THEORY OF SCIENTIFIC FOUNDATION   41 spontaneous creative acts here insofar as these do not follow with necessity from this cultural context, even though they do arise in response to problems posed within it. Thereby it is not always merely a matter of the generation of new axioms and concepts for a theory, but also and more directly a matter of the reexamination of the entire interpretational schema of experience. Thus it is not merely a matter of changing the coordinate systems, the space/time representations, and the fundamental concepts in use, such as mass, force, and acceleration, but also one of reexamining the meaning ascribed to experiments, the interpretation given the instruments, and how seriously one takes confirmations or failures, unity and completeness or the lack of such, in the realm of science. Not only a theory in the narrow sense but moreover the whole theory of science related to this theory is generated in this way. Aristotle and Ptolemy had a completely different conception of the idea of unity or the role of observation from that of Kepler and Galileo; Descartes had a completely different view concerning the essence of confirmation from that of Newton, etc. The radical changes are so comprehensive and so far-reaching that the constant ends of *bon sens* assumed by Duhem, as well as the ontological belief bound up with these, are proved to be mere fictions when set over and against this animated manifold of history.

4.3 The Introduction of Categories
and the Emendation of Duhem's Theory

The critical encounter with Duhem has already led us beyond him. I would now like to attempt to give a systematic foundation to what has hitherto appeared more in the form of a promising observation. In order to accomplish this I will introduce several categories here. In this way I will also give an initial outline for the idea of a historistic theory of science, as this arises out of Duhem's original proposal.

In conjunction with Duhem I begin with the observation set down in the previous chapters that both the construction and judgment of a theory presuppose a series of accepted or established precepts (*Festsetzungen*) which have neither a logical nor a transcendental necessity. However, contrary to Duhem, as already stated, I am of the opinion that these precepts, taken now in a much more radical sense than is allowed for by Duhem's conception of *bon sens*, can only be grounded and understood historically. They are contingent precepts. And because the historical element of a theory resides in these precepts, a historistic theory of science will have to begin by establishing a guiding thread to be used in their systematic disclosure.

Five major groupings of such precepts are discernible:

First, precepts that lead to the procurement of measurement results (precepts concerned with the validity and function of employed instruments and means, etc.). I call this group *instrumental precepts* (*instrumentale Festsetzungen*).

Second, precepts that are used in the formulation of functions or natural laws on the basis of measurement results and observations (for example, limitations placed upon the selection of measurement data, margin of error theories, etc.). These can be called *functional precepts* (*funktionale Festsetzungen*).

Third, precepts that serve as axioms introduced for the deduction of natural laws and by means of which experimental predictions are made with the aid of boundary conditions (*Randbedingungen*). These might be called *axiomatic precepts* (*axiomatische Festsetzungen*).

Fourth, precepts that govern the acceptance or rejection of theories on the basis of experiments. (The following belong to this group: (*a*) precepts that serve as a basis for deciding whether the theoretically deduced predictions agree with the given results of measurements or observations; (*b*) precepts that indicate whether the theory in question is to be rejected or retained in the event of a failure of agreement and, if retained, whether it is to be altered, as well as where such alterations are to be made.) These we can name *judicative precepts* (*judicale Festsetzungen*).

Fifth, precepts that serve prescriptively in the determination of those characteristics which a theory should generally possess (for example, simplicity, a high degree of falsifiability, self-evidence (*Anschaulichkeit*), the satisfaction of certain causal principles or empirical criteria of meaning, and other similar traits). These can be called *normative precepts* (*normative Festsetzungen*).

The above listing makes no claim to being exhaustive.

These five concepts describe those kinds of precepts which are indispensable for the formation, examination, and judgment of theories in physics insofar as these theories are related to measurements—and this applies no matter what the particular content of these various kinds of precepts might be. This holds because whenever a theory of this sort is sought, we *must* decide on the particular form the theory should have and on particular axioms (thus we must decide on normative and axiomatic precepts); at the same time we *must* establish a transposition mechanism linking the theory with the experimental results (hence we *must* devise instrumental, functional, and judicative precepts). However, with regard to particular cases, there is no necessarily valid prescription for *how* we go about all this.

It is therefore a condition of the possibility of a theory of physics that there be particular precepts pertaining to it in such a manner that each of these precepts must fall into at least one of the above-mentioned five concepts, and thus that none of these remains empty. For this reason we can then call these concepts *scientific-theoretical categories* (*wissen-*

schaftstheoretische Kategorien).ᵉ Of course, these categories should not be confused with the Kantian categories, just as those discussed in chapters 2 and 3 should not be confused with Kant's synthetic judgments a priori. The scientific-theoretical categories under discussion here are distinguishable from the transcendental categories in many ways, but most importantly in that they have no necessary validity. This is the case because they relate only to scientific knowledge and not to knowledge in general; furthermore, some of them relate only to scientific knowledge which has come into prominence since the application of instruments of measurement became a basis for this knowledge. To this extent, then, these categories are historical, even if they do possess a peculiar constancy. For example, if one were to view Aristotelian physics as well from this perspective, then the categories would have to be modified. Instead of speaking of instruments, we would have to speak of sense organs and sensory precepts; precepts of induction would take the place of functional precepts (since an explicit notion of function was still lacking); and the other categories could also be reworked accordingly, with varying degrees of modification in their meanings.

As we have already seen in chapter 3, the picture that we draw of nature in physics is indeed dependent upon particular precepts, though this is only codetermined by them and thus retains its empirical characteristics. Let me briefly clarify this with an example for each of the five categories:

One instrumental precept states: The behavior of rigid bodies follows the laws of Euclidean geometry.

A functional precept: From a series of measurement data, a function can be derived by means of the Newtonian interpolation formula.

e. The term *wissenschaftstheoretisch,* translated here and in what follows as "scientific-theoretical," has a rather complex meaning. It pertains to "theories of science." But theories of science are employed and developed both by people who are primarily theorists of science (philosophers of science) and by people who are primarily scientists, in a somewhat narrower sense. Sometimes these distinctions overlap to a great extent (as with Bohr, Einstein, Heisenberg, etc.), whereas sometimes they are separate to a greater degree (as is the case when we compare Popper, Carnap, and Hübner himself, etc., with any number of modern physicists who do not concern themselves primarily with the theory of science as such, but merely employ or embody a particular stance in their research. Hübner uses the term in such a manner that it can, and in his opinion *should,* be applicable to both of these areas. The primary emphasis does indeed lie with the "theory of science" and those people who develop such theories; but it is also meant to refer to anyone who is engaged in significant scientific research. Thus his "scientific-theoretical categories" are meant to be enlightening both for theorists of science and for scientists as such. The same complex meaning also pertains to his use of *Wissenschaftstheoretiker* (scientific-theoretician) and should be held in mind when this term is read. It is for this reason that the hyphenated term has been used in the translation rather than simply rewriting this as "pertaining to the theory of science," since in English the use of "theory of science" might call to mind something which is somewhat more separate from science itself than is the case in Hübner's use of *wissenschaftstheoretisch* or *Wissenschaftstheoretiker.*

An axiomatic precept: All systems of inertia are to be viewed as equal.

A judicative precept: As soon as a theoretically predicted result is not obtained, the theory must be abandoned (the radical principle of falsification).

A normative precept: All theories must be in agreement with a deterministic and hence unlimited principle of causality.

All of these cases deal with decisions concerning the manner in which nature is to be interpreted. Once these decisions are made, their consequences are empirical. If these decisions are changed—and this has occurred in the course of the history of physics in all five categories—then they yield other consequences or results, and this too is an empirical fact. With our precepts we project a framework without which there is no physics. However, the *manner* in which nature presents itself within such a relative framework, *how* it appears within it, this is an empirical fact. 89

4.4 The Importance of the Categories under Discussion for the History of Physics

The five categories given here, or their corresponding historical modifications, place in our hands the means for a systematic investigation of the history of physics in terms of those precepts which are variously operative in it. This history has proved itself exceedingly important for the question of foundations in physics; indeed these categories are that guiding thread by means of which this history can be written.

If we proceed in this way and view matters in this light, then as a result we also no longer find the nonhistorical constants purported by Duhem. In opposition to this, we can now assert that most of what can be subsumed under the categories has changed, just as have the categories themselves, and that this transformation, like the categories themselves and what is subsumed under them, is a function of historical situations—situations which do not exist merely in terms of the history of physics.

The history of physics evinces its fundamental importance for the theory of science in that it frees our vision for the study of the interrelations involved here and offers us illuminating examples of this. It thus presents us with a kind of propaedeutic.

The brief examination of the period leading up to Newton has already indicated that the grounds for all these precepts can lie in various realms outside the immediate province of physics—for instance, in theology, in metaphysics, indeed in the entire cultural situation, including such areas as politics, economics, and technology. 90

Of course, such precepts are often only the direct consequences of other precepts. Indeed, precepts belonging to different categories are not as a rule created independently of one another; rather they betray a hierarchy within which various categories are interchangeable in order of importance. At times, especially when there is a major scientific shift, the axiomatic, judicative, and normative precepts become dominant and thus determinative of the others. At other times it is shown that the instrumental and functional precepts become foundational and draw the axiomatic ones along in their wake. It is especially in this latter case that this relation of dependency often gives the impression that the only kind of reflection at work is that which belongs to physics proper. For in cases where the experiment alone seems to decide matters, the presuppositions involved are especially difficult to perceive. But why then is one set of precepts primary in one instance and another in some other instance? As all of this treats of precepts, one is led—as the previous historical example has already shown—continually beyond these precepts into the realms where they have their immediate origins. And these realms often lie outside physics proper.

Planck's deduction of his laws of radiation and the introduction of his theory of a quantum of action (Planck's constant) appear to me to be a case where the instrumental, functional, and judicative precepts held the upper hand in order of priority. In opposition to this, as was shown in chapter 2, the relation is entirely different in the later development of quantum mechanics, where the axioms were re-formed by the physicists of de Broglie's school in accordance with the expressed goal of turning back toward a deterministic interpretation of quantum mechanics.

In addition, we might well call to mind here the various reasons which have been given in support of the normative rule to always choose that which is simplest—whatever the more exact intentions of this might be. These reasons can be generally summarized as follows: Nature should correspond to divine wisdom, and so it should be simple; or simplicity is necessary because simplicity would allow man to reach the desired goal of physics more easily, namely, the mastery of nature; or because a simpler physics is more beautiful, etc. This running example also shows that general determinations of goals, which for their part are again only to be understood historically, play a decisive role in the selection of precepts.

Today there is a widespread tendency to view the conditions of the genesis of a theory as psychological and accordingly as something which is unimportant for the theory of science as such. It is asserted that the theory of science is only supposed to concern itself with scientific theories, in whatever way these might have occurred, and with the results which follow from these theories or are directly tied to them.

The first point to be raised in opposition to this view is that here the subjective *act* of the genesis of a theory (something like the conditions of an inspiration) is confused with the derivation or explanation of its foundations—and thereby a psychological phenomenon is confused with a historical one. Second, we must realize that the same kinds of motives, sometimes even the same motives, which are determinative for the founding of a theory (as, for example, the motives operative in the formulation of its axioms) also play a decisive role in the testing of a theory. The set of judicative precepts shows this pointedly. When someone, as for example Popper, brushes aside the question of the founding of axioms as irrelevant for the theory of science because this problem is thought to be only of a psychological nature and in exchange for this expressly sets up criteria of falsification,[14] in doing so he overlooks that to every falsification there belong (*a*) the acceptance of certain basic statements and (*b*) the decision, based on this acceptance, to actually reject or accept the theory. And this acceptance of basic statements and this decision to reject or accept a theory, along with the motivating reasons for both of these, are analogous to those reasons which have led to the formulation of the axioms, since none of these matters is necessarily determined. Whoever is not satisfied with the blind acceptance of certain criteria and rules of falsification—and who could be?—will then have to inquire into the goals and objectives lying behind these criteria and rules; and in doing this one is then already, as we have just seen, far beyond the realm which is commonly thought to pertain to the theory of science alone. It is impossible to admit the existence of determinative historical motives as operative in the initial stages of a theory (thus for the formulation of the axioms) and then to disregard these again in the consideration of the final stage of the theory (thus for the procedures by which it is tested). This point will become a topic for subsequent chapters, especially chapter 10.

4.5 The Propaedeutic Significance of the History of Science for the Theory of Science

Whether it be then a matter of the precepts themselves, of their relations to one another, or more directly of the changes in them, we find that the work done by the historian of science supplements that of the scientific theoretician. The material brought to light through such research can itself serve as a starting point for reflections on the theory of science, and it is for this reason, as I have already stated, that such research has a propaedeutic meaning for the theory of science. In the following four points I now wish to illustrate how this is the case in a more precise sense and how, beginning from this perspective, new areas of research within the theory of science are delimited.

First: On the basis of historical facts, a typology can be developed for the above precepts and their relations, as well as for the grounds for these precepts and the changes which take place in them.

The meaning of this statement is shown by the examples introduced earlier concerning particular precepts and the goals or purposes connected with them. I mentioned as possible normative concepts for a theory (among others) simplicity, a high degree of falsifiability, self-evidentness, satisfaction of certain causal principles; and in addition the theological, pragmatic, and aesthetic ends which might underlie the former. However, a typology developed in terms of such or similar material should not serve merely as a categorizing survey of these things, of what was and is, but rather should be taken as a mere starting point for a further sounding-out of scientific-theoretical possibilities. This systematic treatment would have as its purpose the classification of the various presuppositions pertaining to the theory of science in use at given times, the warehousing of such knowledge in anticipation of possible further application, and finally the facilitating of the formulation of new presuppositions.

Second: With the aid of this historical information, the historical origins of applied or formulated scientific-theoretical rules, methods, and principles can be elicited and held in mind.

94   Insight into this historical conditioning prevents the progressive degeneration which so often accompanies the acceptance of scientific positions—a degeneration which moves first to the level where the position is accepted uncritically, then to a level where it is thought to be somehow self-evident, ending finally in a stage where all questionability has disappeared. In this way historical awareness possesses a critical function. Over and over again it tracks down origins that have only contingent meanings, and thus lack necessity or compelling grounds. And it is precisely for this reason that historical consciousness can reject such positions. Therefore the theory of science must not be allowed to limit itself to the function of merely uncovering historical conditions and stipulating these as constants; over and above this it must grasp that only such historical research can serve as a sufficient basis for an adequate critique. Various scientific-theoretical foundations exhibit themselves only through the study of the history of science; hence it is here that the prerequisites for a correctly founded critique present themselves. In this way we do justice to the situational complex (*Situationsgebundenheit*) on the one hand while, on the other, we are enabled to critique this precisely because we are now acquainted with it.

Third: Historical material is to be employed as a standard against which to judge the scope, validity, and applicability of the methods, principles, postulates, etc., that have been worked out by scientific theoreticians. It is indeed very informative to see how little the classical physicists held to certain doctrines of modern scientific theoreticians; in fact, had they done so, they could scarcely have argued for their own theories. (Cf. esp. chapters 5, 6, 9, and 10.)

Fourth: In all those places where the categorial presuppositions stem from historical realms which do not genuinely belong to physics—and sooner or later one always encounters this—the discussion pertaining to the theory of science must also be extended so as to include these extra-scientific realms.

Hence it is not enough to take superficial notice of the fact that certain   95
precepts stem from (for example) theological, pragmatic, or aesthetic
objectives and purposes; rather these purposes must themselves become
the expressed object of reflection and critique. As a rule this leads far
beyond the theory of science, considered in the narrower sense, into the
traditional realms of philosophy. But this expansion beyond narrow
boundaries is unavoidable if we wish to discuss the justifications for historical precepts.

All of the tasks enumerated above serve the practical purpose of clarifying the scientific-theoretical presuppositions used in the realm of the
exact natural sciences, when viewed with respect to their historical foundations and limits. Thus they serve to create a critical distance from these
presuppositions. This in turn then serves as an aid for those cases where
it might become necessary to employ other already existing presuppositions within a particular situation as well as in the formulation of new
presuppositions.

We might further illustrate this as follows: The typology, developed in
accordance with the guiding thread of the categories, should help the
researcher to initiate reflections concerning the conditions of his activity,
to delimit these in terms of concepts, and to recognize that these conditions, while certainly different from others, should nonetheless be viewed,
first and foremost, as only *one* possibility among others. In this way he
first becomes aware of what he is doing and is also freed to take up a
critical view of it. By taking a further step, he will recognize the historical
conditionality and relativity of his scientific-theoretical premises; along
with this he will also become aware of the question this poses, namely,
whether he still desires to view these conditions and this relation as binding. Through this attempt at a more profound justification of his presuppositional decisions, he will ascertain that his own area of concentration   96
leads back principally to the traditional realms of philosophy—and we
can observe this with all great researchers in that they did not acquiesce
to the demands and constraints of a naive realm of specialization. All of
these reflections will then finally place him in a position to make a well-
grounded choice as to whether to stand by his presuppositional decisions
or take up others stored in the typology, and perhaps with their help
formulate things anew. When I say "grounded" (*gegründet*), I do not
mean—as is obvious from all that has been said here—some kind of
absolute basis. Rather I mean only that he has then at least seen and
thought about all of those components which can be viewed as related to
such a foundation. A discussion of the entire realm antecedent to a theory
can never have an absolute end; but *whatever* is to be discussed in this
connection results from the points listed above.

Thus the well-known and partially unsolved questions generated by
historical thought, questions which seemed previously to have only been

touched upon by the human sciences (*Geisteswissenschaften*), now crop up as well in the very lap of the natural sciences. The boundaries between the two can no longer be drawn in the old manner and with the old rigor. Chapter 13 will be concerned especially with this question. But first of all the previously developed thoughts will be explicated and plumbed more deeply by means of several pertinent historical examples.

CRITIQUE OF THE AHISTORICAL THEORIES OF SCIENCE OF POPPER AND CARNAP USING THE EXAMPLE OF KEPLER'S "ASTRONOMIA NOVA"

The theories expounded by Kepler in the *Astronomia Nova,* insofar as   97
we will be concerned with them here, take their impetus from the attempt
to determine the orbit of Mars. After years of the most strenuous effort,
Kepler finally concluded that his initial attempt was doomed to failure.
This occurred after he had been forced to concede that between the
calculated assessment based on his hypothesis and that based on the
observations of Tycho Brahe there was a difference of eight minutes of
arc. He wrote:

> To us, to whom divine beneficence made a gift of the most diligent observer
> in the person of Tycho Brahe, through whose observations was brought to
> light an error in the Ptolemaic calculation of the orbit of Mars in the amount
> of eight minutes, it seems fitting to accept this favor of God with a thankful
> mind and to use it. In this vein we will work in order finally to ascertain (on
> the basis of the reasons for the incorrectness of the suppositions that were
> made) the correct form of celestial motions. In what follows I wish to lead
> the way for others in my own manner. . . . These eight minutes alone have
> thus shown the way for the reformation of the whole of astronomy; they have
> become the material for a great part of this work.[1]

Such a statement is quite familiar to us today, and accordingly only his   98
fervor seems out of place. Here, so one says, is the origin of modern
natural science in that Kepler called upon data. This is indeed so; however,
all too often we overlook the fact that the carelessness with which such
data were handled earlier, the indifference with which even greater dis-
crepancies were accepted than those which occasioned Kepler's rejection
of his hypothesis concerning the orbit of Mars[2]—for which reason he
called this a *hypothesis vicaria* (vicarious, or provisional, hypothesis)—
can in no way be ascribed to a lower niveau of science or to personal
insufficiences among the scholars engaged in scientific inquiry. This earlier
attitude is actually much more closely tied to that theory which molded
the Ptolemaic age. The fundamental formula of this theory, the so-called
Platonic axiom, which stated that the heavenly bodies moved in circles

51

with uniform angular velocity, was grounded in an extensive metaphysics that dealt with the difference between the celestial and the earthly orders of being, the perfect and the imperfect, the higher and the lower. This earlier theory had as its expressed purpose "to save or preserve the phenomena" (σώζειν τὰ φαινόμενα); and thus it had to bring order to the confusion of appearances by means of an application of the ruling metaphysics. Where this did not succeed, the explanation was already waiting in the generally accepted premise: Who would dare to bestow blind trust upon perception?—least of all where this perception was of objects so removed and sublime as those relating to the heavens. Such perception might be more or less valid for the sublunar realm; but it had no final competence to judge the movements of the heavenly bodies. It would be naive to see Kepler's radical rejection of this attitude, coming as it does at this early time, as already representative of the victory of reason and science as these are understood today. In truth Kepler shows himself to be guided only by other metaphysical thoughts than those of his opponents. Behind Kepler's words cited in the previous passage stand the humanistic-theological, fundamental assumptions of Copernicus already discussed in the previous chapter, namely, the assumptions that the Creation has a structure well suited to human understanding; that consequently spirit or mind (*Geist*) and perception should not come into conflict; that the distinction between lower and higher orders is meant to be overcome, the Earth is to be viewed as a star among stars and has its part to play in their dance of circles; that the universe has to be construed according to principles of simplicity, etc.[3] But the Copernican system, with all its humanistic-theological presuppositions derived from the spirit of the Renaissance, was in truth much more poorly grounded at the time than the Ptolemaic. In order to support this system, and this too has already been mentioned, one merely turned the tables on the Aristotelian thinker, positing ad hoc theological arguments in opposition to theological arguments and metaphysical arguments counter to other metaphysical arguments. There was no single, directly compelling ground capable of validating the new system, especially since the rotation of the Earth had to remain an unresolved riddle until the principle of inertia—first formulated by Newton—could explain why there was no perception of this fact.[4]

Kepler's decision to follow Copernicus—and thereby to accept perception and observational data as the final scientific court of appeals—is therefore initially more comparable to a spontaneous act than to the necessary result of rational reflection, however this latter notion might be understood; and yet Kepler did arise out of a cultural context which had long been developing counter to the Ptolemaic system.

5.1 A Scientific-Theoretical (wissenschaftstheoretische)[a] Analysis of Kepler's "Astronomia Nova"

Kepler's initial failure to calculate the orbit of Mars led him to the idea   101
of concerning himself next with the orbit of the Earth.[5]

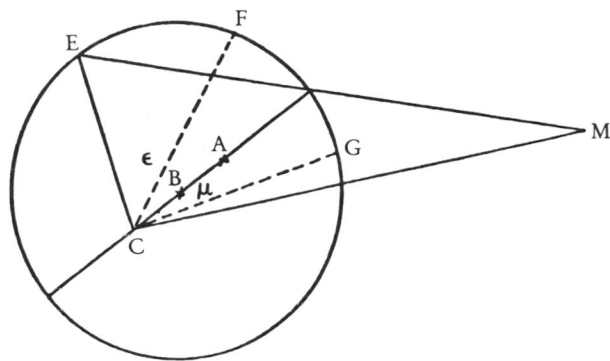

Figure 1

To this end and with the help of a theory developed by Tycho Brahe, he determined the heliocentric position of Mars (M in figure 1) and of the Earth (E) with respect to a given time point (T). The angles ϵ and μ, formed by the respective radius vectors and the diameter upon which the segment AC lies, serve to illustrate how this position is determined. One must keep in mind, however, that figure 1 does not present these orbital relations as they were taken from Tycho Brahe, but rather as they were first calculated by Kepler with the help of Tycho's data.

Obviously heliocentric means here related to point C; but this does not   102
correspond, as Tycho had already known, with the Sun (A in figure 1), nor as might have been supposed, with the center point of the Earth's orbit (B)—though this view did arise later on. Now, by incorporating the position of Mars relative to the position of the Earth (geocentric longitude), Kepler could calculate the parallax EMC which follows from this view, and the angle CEM.[6] Accordingly, he arrived at the relative distance of the Earth from point C (employing the law of sines) by means of the equation

$$CE = CM \, \frac{\sin EMC}{\sin CEM},$$

a. For a statement of the meaning of *wissenschaftstheoretisch* (scientific-theoretical), see, tr. note e, chapter 4.

in which he posited CM = 100,000. Then Kepler chose a second time point t′, when Mars was once again in the same position after having completed a full orbit, while the Earth, for reason of its own movement, was in another position (F in figure 1). By employing the same method, Kepler again calculated the distance of the Earth from C (CF). Finally he chose a third corresponding time point t″ (as well as a fourth, which we omit from our treatment here); accordingly he arrived at a third position of the Earth (G), with the distance (CG). The result was that all three of the calculated distances differed from one another. From this he concluded that C could not be (as presupposed) the center point of that circle upon whose periphery the three positions of the Earth lay. More appropriately, C had to represent the equant point (*punctum aequans*), hence that point around which the Earth moved with a constant angular velocity, since there lay a full Martian year between each of the three positions, yet the angles formed on the one hand by CE and Cf and on the other by CF and CG were equal.

103

Now, Kepler still wanted to calculate the distances of the equant point C and the Sun A from the orbital center point B, and also to determine the position of the apsis line (thus, the position of the diameter upon which A, B, and C lie). However, AB could be ascertained only if the actual heliocentric longitude of Mars relative to A was known (and not the previously termed "heliocentric longitude" of Mars, which in truth was relative to C). Therefore here it was not possible for Kepler to support himself on the basis of Tycho's theories; so he boldly took up again his own, previously rejected *hypothesis vicaria* and now sought to balance out the error in it through a crude process of approximation. The results of this can be stated as follow: The Earth, as well as Mars, moves in a circular orbit with a divided eccentricity (*geteilte Exzentrizität*),[b] namely, the two eccentric points, the equant point C and the Sun A in figure 1, which lie on the apsis line equally distant from, and on opposite sides of, the center point of the circle.

But in the final analysis what was this discovery based on? It was based on two theories that were questionable even to Kepler himself: (1) Tycho's theory (involving the heliocentric positions of Mars and the Earth) and (2) Kepler's own *hypothesis vicaria,* which he had rejected earlier with such fervor. In addition, it incorporated a crude process of approximation; it accepted the classical-philosophical assumption of the circular movement of the heavenly bodies; and finally it rested on Tycho's data, which were held to be almost infallible.

b. The expression *geteilte Exzentricität,* translated here as "divided eccentricity," is a fabrication for which there is no standard meaning in either German or English. The intended meaning is given in the text. But it is important to understand that this is a crucial step in the movement which eventually led Kepler to the description of planetary orbits as ellipses— geometrical structures having two foci.

But these dogmatic and questionable presuppositions did not prevent Kepler from taking yet a further bold step, in which his divergence from Ptolemy as well as Copernicus is already quite clearly shown. Namely, 104 he gave up his attempt to construe the equant circle—a task handed down to him by tradition, which utilized the *punctum aequans*—and in place of this he sought to find a law in the nonuniformity of the Earth's orbital velocity as it traveled around the Sun. Thereby he found—again by means of approximation—that the velocity of the Earth relative to its perihelion and its aphelion stands in an inverse relation to the distance of the planet from the Sun at these points. And this minimum of empirical data was already enough for him to go ahead immediately and extrapolate from it to all points on the orbital path and from thence to all planets, such that he formulated the following law:

1. All planets move in circular orbits with a divided eccentricity, where the Sun occupies one of the eccentric points.
2. The velocity of a planet is inversely proportional to its distance from the Sun.

This second assertion is the so-called law of the radii.

Not only is the speculative nature of this law noteworthy, but also the very fact that Kepler ever sought such a law at all and abandoned the attempt to construct the equant circle. In doing this he had already given up a part of the Platonic axiom, namely, that the planets move with uniform angular velocity. And what guided him in this was his mystical view of the Sun. Fictive points around which the heavenly bodies are supposed to move were phantasms for him. He had already been disturbed by the fact that in the Copernican system the Sun did not truly stand at the center point (wherefore it could not be called "heliocentric" in a strict sense)[7] and that accordingly in that system it played only the auxiliary role of a 105 source of light. For Kepler the Sun was the holy center of the universe and the inner-worldly expression of God the Father. Thus the force which whirled the planets around the Sun could only come from the Sun (he related this force to the Holy Ghost and the fixed stars to the Son). So it was essential to seek out this force; hence the movements of the planets had to be investigated in relation to the Sun and not in relation to some kind of imaginary point in space.

Thus it was only this passion for heliocentrism that enabled Kepler to seek out and find something like the law of the radii, and only his unshakable conviction, grounded in Renaissance humanism, that the principles by which the universe was constructed were intelligible gave him the courage to trust his bold extrapolations and to see them as proof. It was then in the spirit of this philosophy that he moved consistently forward when he undertook what was for the Aristotelians the incredibly daring task of relating the law of the radii first to the principle of the lever and then, later, to Gilbert's theory of magnets—thus relating celestial and earthly movements. All of this further entailed the view that the universe was no longer to be viewed "instar divine animalis" (like a divine life form), but rather "instar horologii" (like a clockwork).[8] Finally, in his

hypothesis concerning the causes of planetary movements, which one might view as a primitive form of Newton's gravitational theory, he supported himself again on the basis of Aristotle in that he separated rest and motion absolutely (if not for the rotational force generated by the Sun, the planets, he thought, owing to their natural inertia, would stand still). Thus he actually obstructed the way leading to the principle of inertia and thereby—as we know today—to the most important argument in defense of the Copernican idea.

After these reflections on astronomical mechanics, he returned to the theory concerning the movement of Mars. Let us now look at figure 2.

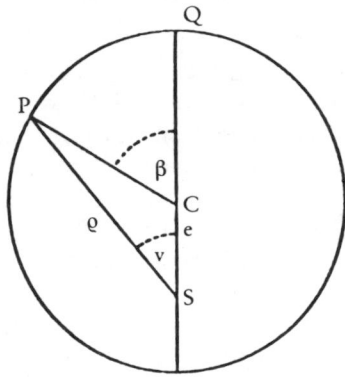

Figure 2

According to the law of the radii, the velocity of a planet at point P on its orbit with center point C is inversely proportional to its distance $\rho = $ PS from the Sun S; consequently the time expended in this segment is proportional to PS. But how might this proportionality be precisely captured in a formula? It appeared impossible to formulate a direct relation between the time and the radius. But then Kepler remembered the so-called Archimedean theorem, which expressed a relation of areas and radii for circles. This theorem stated that the area of a circular sector QCP can be

viewed in the limiting case as the sum of infinitely many, infinitely narrowing triangles for which the altitude is the radius of the circle. This led Kepler to the idea of relating the elapsed time for the movement QP not directly to the radii, but rather to areas understood as sums of radii. Thus without further thought he appropriated the Archimedean theorem for the sector QSP, and thereby obtained a rather questionable means by which to express the time elapsed in the arc QP by utilizing the area QSP, and thus at least an indirect means of formulating the relation between the time and the radius. With this he then formulated the following relation:

(1)
$$\frac{t}{T} \approx \frac{1/2\ \beta\ +\ 1/2\ e\ \sin\ \beta}{\pi}\ ,$$

where t denotes the elapsed time for the arc QP and T represents the time of a complete orbit. If we posit r = 1, then the area QCP = 1/2 β, the area CSP = 1/2 e sin β, and π is the area of the circle.

From (1) it then follows that

(2)
$$2\pi\ \frac{t}{T} \approx \beta\ +\ e\ \sin\ \beta.$$

β could then be calculated if t were known (although with the methods at Kepler's disposal this could be only roughly approximated).

Thus the distance between a planet and the Sun can be determined by the equation

(3)
$$\rho = \sqrt{1\ +\ e^2\ +\ 2e\ \cos\ \beta}\ ,$$

which results from figure 2 under the use of the cosine law. Finally this yields the equation

(4)
$$\rho\ \cos\ v = e\ +\ \cos\ \beta\ ,$$

which, by utilizing the simple cosine relation, yields v, and thereby the position of the planet at a given time t.

108

These considerations make use of the following: first, the law of the radii, whereby a relation between time and the radius is asserted; second, the transferral of the so-called Archimedean theorem—by means of which the area of a circular sector is calculated in the limiting case by means of the radius—to something which is completely different from the sector of a circle (namely, QSP). In this way the relation between time and the radius is transformed into one between time and the area of the circle. But the law of the radii was hardly well supported on empirical grounds and the transferral mentioned above was mathematically unsound. Both of which points were well known to Kepler. To this can be added the fact that the eccentricity e also figures into equations (1) through (4); and this he only obtained by means of his *hypothesis vicaria,* which had been rejected earlier.

In this stage of his development as well, then, Kepler shows himself to be little concerned with exact and adequate empirical, mathematical, or theoretical foundations, although, according to the passage cited at the beginning of the chapter, such foundations are anticipated as imminent by him. Thus it becomes much less surprising to see how, with a minimum of empirical evidence, he finally rejects the remaining part of the so-called Platonic axiom—namely, the assumption of the circular form of planetary orbits—as he had already given up half of this framework, i.e. the assumption of uniformity in the angular velocities of the planets.

He arrived at this stage in his renewed attempt to determine the orbit of Mars. In this he initially availed himself of the above-described method, which he had already used in the calculation of the Earth's orbit. But whereas he had proceeded in the first case by means of the comparison of several positions of the Earth relative to the position of Mars, here he began with three positions of Mars relative to the same position of the Earth. Thereby he obtained three distances of Mars from the Sun as well as the angles formed by the corresponding radius vectors. By means of tedious, though in themselves rather simple, trigonometric calculations, he ground out the position of the apsis line and the amount of the Sun's eccentricity in three separate instances. Thus he could not fail to see that in each instance he arrived at different results. From this he concluded that the orbit of Mars could not be a circle.

He made this revolutionary decision on the basis of the same bold presuppositions he had already assented to in the calculation of the Earth's orbit. Here he used the same method as there; here again he rested his case on the results already won from Tycho's theory, the *hypothesis vicaria,* and the belief in Tycho's data. All of this remained, as before, the shaky ground upon which he stood.

So even in the last stage of his journey, where he arrived at the conviction that the orbits of the heavenly bodies were ellipses, this speculative spirit, as we shall see, will remain true to form. Let us look at figure 3.

First, following the principle of simplicity, Kepler postulated that the deviation of the orbit of Mars from a circular orbit could be given by the relation $b = 1 - e^2$, where 1 is the radius, e the eccentricity of the Sun, and b the axis of the actual orbit. Later on he came up with the idea of positing $b = 1 - (e^2/2)$.

But one day he made a discovery, which is illustrated in figure 4 (representing the orbit of Mars). He noticed that (see figure 4)

(5)
$$\frac{P_1C}{P_1S} = \frac{1}{\cos\phi} = 1.00429 .$$

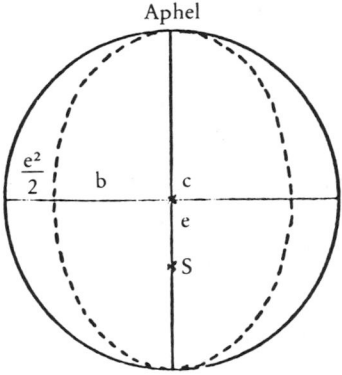

Aphel

Figure 3

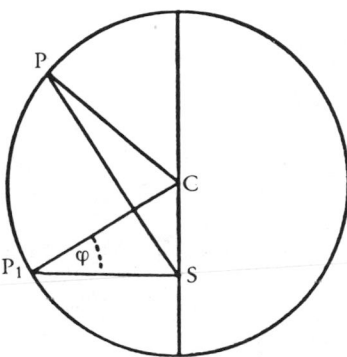

Figure 4

Here ϕ is the greatest angle formed by the convergence of the segment P_1S (planet–Sun) and P_1C (planet–center point of the circle). If we then merely substitute the assumed value of b in the calculations, it follows that

$$\frac{r}{b} = \frac{1}{1 - (e^2/2)},$$

and because $e \ll 1$

$$\frac{1}{1-(e^2/2)} \approx 1 + \frac{e^2}{2};$$

111 however, $1 + (e^2/2)$ has the value 1.00429, which agrees with the calculated result in (5).

"When I saw this," wrote Kepler, "it was as if I had been awakened out of my sleep and saw a new light."[9]

This relation, though merely approximate and valid only because of the minuteness of e, immediately spurred him on to a new speculation, illustrated in figure 5.

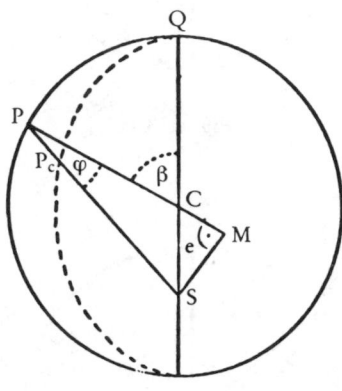

Figure 5

He postulated that in figure 5 the following relation, analogous to equation (5), might hold:

$$\frac{SP_c}{SP} \approx \cos \phi .$$

Otherwise expressed, he postulated that the relation of the distance between the Sun and the planet on the "true" orbit and the distance between the Sun and the planet on the "fictional" orbit might be analogous to the relation r : b in figure 3.

112   Now with r = 1, by using figure 5 we arrive at

$$SP \cos \phi = PM,$$
$$PM = 1 + e \cos \beta.$$

From this it follows that the planetary orbits are expressed by the formula

(6) $$SP_c \approx 1 + e \cos \beta.$$

After arduous labor—"paene usque ad insaniam" Kepler then estab-

CHAPTER 5

lished (and here it must be remembered that the mathematics available to him was still rather primitive) that equation (6) expressed the formula for an ellipse (this being only an approximation as well).

Thus, here we once again encounter suppositions, speculations, and crude methods of approximation; and what is more, the manner for checking equation (6) involves the comparison of the value for SP_c with that which had been obtained by means of Kepler's methods for the determination of distance, which have already been critically analyzed.

In conclusion let us look at yet another of Kepler's steps, as illustrated in figure 6.

Corresponding to formula (1), the following relation should then also be valid here: 113

$$(7) \qquad \frac{t}{T} \approx \frac{SQP_c}{\pi b} \, .$$

In other words, the time t which is needed for the planet to complete the elliptical arc QP_c should be related to the total orbital time T as the area of SQP_c is related to the total area of the ellipse, where b is the radius of the minor axis and 1 is posited as the radius of the major axis. Next Kepler assumed the following, which is analogous to what has already been stated with respect to figures 4 and 5:

$$(8) \qquad \frac{SQA}{SQP_c} \approx \frac{1}{b} \, .$$

However, according to (1),

$$SQA = 1/2 \, (e \sin \beta + \beta) \, .$$

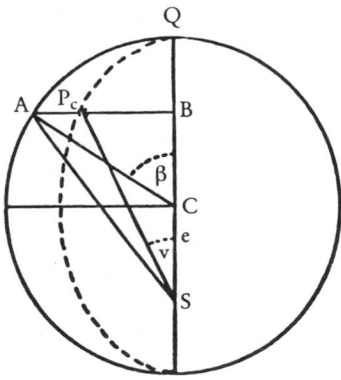

Figure 6

And if this is substituted in (8) and (7), then by simple calculations we finally arrive at

$$\frac{t}{T} \approx \frac{e \sin \beta + \beta}{2\pi} \ .$$

The decisive step in this deduction—namely, taking equation (7) as a beginning point—presents us with nothing other than a revised and highly questionable application of the Archimedean theorem: In this instance, it is applied to a sector of the ellipse, the tip of which is the Sun, which is now taken as one of the foci of the ellipse.

Thus we can formulate Kepler's first two laws as follows (see equation [6]):

(9) $$2\pi\frac{t}{T} \approx \beta + e \sin \beta \ ,$$

(10) $$SP_c \approx 1 + e \cos \beta \ .$$

114   Equation (10) states that a planet moves in an ellipse, one of whose foci is the Sun. Equation (9) states that in equal times the radius Sun–planet sweeps out equal areas.

With respect to Mars, the central concern and point of origin for all these thoughts, this means that β and e in equation (9) were only to be determined by means of the previously rejected (even if now improved) *hypothesis vicaria*. This hypothesis was thus used both in the calculation of SP_c and in the method by which the ensuing result was validated and checked (incorporating, as it does, the determination of three positions of Mars relative to the same position of the Earth).

So this is what the foundation of Kepler's first two laws actually looks like, a foundation which even today is often viewed as if it had arisen on the basis of experience alone.

If, in addition, we then compare Kepler's theory with that of Ptolemy, the latter theory does not fare badly at all. In the first place, because of the minuteness of the orbital eccentricities of the planets, the Ptolemaic system describes planetary movements with almost the same accuracy as Kepler's theory (in the case of Mercury, both theories alike present us with an *enfant terrible*). Second, there was a lucid philosophical foundation for the Platonic axiom whereas for Kepler the elliptical form of planetary orbits had to remain an enigma. His attempt to deduce these from characteristic motions of the planets remained totally unsatisfactory. Third, the same thing must be said about his efforts to overcome the accepted Aristotelian arguments against the rotation of the Earth. These
115   efforts have the character of typical ad hoc hypotheses.[10] Hence the si-

lence with which his *Astronomia Nova* was received is not at all surprising.

This analysis of the methods and evidence employed in the *Astronomia Nova* leads us, as will shortly be shown, to the following certainty: If Kepler had held to the recognized doctrines pertaining to the theory of science of our times, he would have had to reject both of his inestimably meaningful laws. This can be illustrated by means of two relevant examples: the methodology of Popper and Lakatos, and Carnap's inductive logic.

5.2 Kepler's "Astronomia Nova" in the Light of the Philosophy of Science of Popper and Lakatos

Popper's fundamental methodological postulate states that a scientific theory must be falsifiable. But if such a falsification occurs, we are not permitted to avert the breakdown of the theory by means of ad hoc hypotheses or other assumptions. He writes: "If this decision is positive, that is, if the singular conclusions turn out to be acceptable, or *verified,* then the theory has, for the time being, passed its test: we have found no reason to discard it. But if the decision is negative, or in other words, if the conclusions have been *falsified,* then their falsification also falsifies the theory from which they were logically deduced."[11]

What does a "decision" (*Entscheidung*) mean here? It means that so-called basic statements (by which Popper understands singular existential statements something like: in such and such a space-time region, this or that exists)[12] either do or do not contradict the theory. But if only certain solitary basic statements contradict the theory, we do not have a sufficient basis for declaring the theory falsified. "We shall take it as falsified only if we discover a *reproducible effect* which refutes the theory. In other words, we only accept the falsification if a low-level empirical hypothesis which describes such an effect is proposed and corroborated. This kind of hypothesis may be called a *falsifying hypothesis.*"[13] For an example of this, Popper uses the statement that a family of white ravens lives in the New York Zoo;[14] whereby the general notion that all ravens are black would be refuted. However, he adds: "In most cases we have, before falsifying an hypothesis, another one up our sleeves; for the falsifying experiment is usually a *crucial experiment* [*experimentum crucis*] designed to decide between the two."[15] Thus we find here that the effect which should falsify is deduced from another hypothesis which is already on hand. To be sure, decisions of such a kind could always be fundamentally and repeatedly called back into question along with the same basic statements upon which they rest; but for the sake of practicality we generally stop at some point and stand by the attested position. For this reason Popper also introduces the following rule: "that we shall not continue to accord a positive degree of corroboration to a theory which has been falsified by an intersubjectively testable experiment based upon a falsifying hypothesis."[16]

118 Kepler, however, did exactly the opposite when he utilized the results of previously falsified theories for the construction of his own theories and then followed this by validating his own theories with the aid of such falsified theories. Moreover, he clearly runs counter to Popper's methodology in yet another manner. The absence of any appearance which might indicate the rotation of the Earth was viewed at the time as a falsification for any form of heliocentrism—and thus for Kepler's as well. In order to get around this falsification, he attempted what Popper decidedly rejects,[17] namely, to save his theory by means of an ad hoc hypothesis and moreover one which was as highly questionable as his astrodynamics. Had he treated matters in the manner prescribed by Popper, he would have had to discard his theory, and this, then, "once and for all."[c] Popper believes that Kepler's success was in part due to the circumstance "that the circle-hypothesis with which he started was relatively easy to falsify."[18] He is correct insofar as he relates the phrase "relatively easy" to the notion that the circular orbit hypothesis is "three dimensional" ("for its falsification at least four singular statements of the field are necessary, corresponding to four points of its graphic representation"),[19] whereas the elliptical hypothesis is "five dimensional" ("since for its falsification at least six singular statements are necessary, corresponding to six points of the graph").[20] But this view tends to veil the problem of exactly how questionable the falsification of the circular hypothesis actually was—a falsification which, as indicated, rested on rather questionable premises.

119 Therefore the example of Kepler reveals not only how difficult it can be to recognize basic statements that falsify (a difficulty which in my estimation Popper has not sufficiently considered),[21] but also that it may not be at all advantageous for science to permit the rejection of a theory in every instance where such a falsification might be recognizable.[22]

120 Thus far Kepler's methodology has only been compared with what is today known as classical Popperianism. But in the final analysis nothing changes when we take into consideration the improvements Lakatos has proposed in recent years.

In his opinion there is a universal rule which must be employed in order to determine whether a series of theories is progressive. (And here he is absolutely correct in speaking of a "series" rather than single theories, since in fact every theory is linked to other, different theories.) He writes:

c. The expression "once and for all" (*ein für allemal*) appears in the original German text, which Hübner cites (Popper, *Logik der Forschung*, 2d ed. [Tübingen, 1966], p. 213). But the English translation, which Popper himself made, obscures the reference, which would be quite obvious in German. The German text in question runs: "daß wir einer durch intersubjectiv nachprüfbare Experimente . . . falsifizierten Theorie ein für allemal keinen positiven Bewährungswert mehr zuschreiben wollen."

Let us say that such a series of theories is *theoretically progressive* . . . if each new theory has some excess empirical content over its predecessor, that is, if it predicts some novel, hitherto unexpected fact. Let us say that a theoretically progressive series of theories is also *empirically progressive* . . . if some of this excess empirical content is also corroborated, that is, if each new theory leads us to the actual discovery of some *new fact*. Finally, let us call a problem shift *progressive* if it is both theoretically and empirically progressive, and *degenerating* if it is not.[23]

Here again we must maintain that Kepler would have had to reject his own theory if he had followed Lakatos's rule.

Kepler in his own opinion could indeed predict certain new, hitherto unknown facts; on the other hand, he was *not* able to explain far more facts which were in complete accord with Ptolemaic astronomy and Aristotelian physics. To this latter group belong especially those phenomena that, owing to the still undeveloped principle of inertia, spoke against the rotation of the Earth. Therefore, in comparison with those theories which preceded his own, we cannot say that Kepler's theory had an "excess empirical content."

But this is not all. Even the confirmation of the facts predicted by Kepler was extraordinarily questionable, as we indicated earlier. Thus we have seen, for instance, that Kepler needed the *hypothesis vicaria* for the calculation of the orbit of Mars and that he then checked the ensuing results by methods which also rested on this hypothesis. In addition, Kepler was well aware of these deficiencies, and accordingly had recourse to yet further assumptions of a more metaphysical and theological nature—a point to which we will return shortly. Can Lakatos's rule be of any help when we want to decide whether all these presuppositions are acceptable?

Obviously the expression "prediction of a fact" is not as clear and simple as Lakatos thinks. Can we view every prediction as a theoretical advance, even when the reasons for these predictions are extremely daring, questionable, or even utterly foolish? Was it not precisely *this* which was questionable in Kepler's case, that is, whether his predictions were acceptable at all, especially since metaphysical and theological considerations entered into them? For the very same reasons, was it not also questionable whether Kepler's empirical tests and confirmation procedures could be accepted? In truth, then, the problem is not whether predictions are made and facts examined, but rather whether the *reasons* for these predictions and the *presuppositions* for these testing procedures are clear and intelligible. But Lakatos's rule says nothing about this.

Let us represent the case as if Lakatos were a Grand Inquisitor at the time of Kepler, charged with the task of overseeing and controlling the progress of science in accordance with his own rule. Assume that he examined Kepler and that the following dialogue took place:

121

122

LAKATOS Can you produce an empirical content which surpasses that of your predecessor's theories?

KEPLER I can indeed indicate something of this, but Ptolemy and Aristotle surpass me in this respect many times over.

LAKATOS Can you predict some new facts?

KEPLER Indeed I can, if you share my reasons for these predictions, and, further, if you accept my presuppositions thereto, presuppositions which one must hold in order to confirm these facts.

LAKATOS What presuppositions have you made?

KEPLER Rather questionable ones, as long as they are viewed only in terms of the purely astronomical realm.

LAKATOS Anathema.

KEPLER Please allow me a final word. I have made two presuppositions that I hold to be of decisive importance and in which I believe wholeheartedly. The first is that Copernicus must be correct, because his description of the world is so much simpler in its essentials than are others and because it is consequently in accord with the mind of man (*menschlichen Geist*) and Divine Justice. The second is that it is contradictory to view the Earth as the center point of the universe and simultaneously as the place of sin. Therefore I believe that the Sun is the star around which all others rotate. And by presupposing *this*, everything else, no matter how questionable it may be in its own right, comes to make sense rationally.

123

LAKATOS None of this has any scientific value. So once again: Anathema.

So poor Kepler would have had to recant his own theory if he had followed Lakatos's rules.[24]

5.3 Kepler's "Astronomia Nova" and Carnap's Inductive Logic

Now let us consider Kepler's theory in the light of Carnap's theory of inductive logic (*Induktionslogik*).[d] The aim of this logic is to determine the degree of confirmation pertaining to a hypothesis (h) as this is grounded in the pertinent givens (e). Therefore its elementary expression is of the form

$$c(h,e) = r,$$

where c represents the degree of confirmation and r has a numerical value between 0 and 1.

Now, to be sure this system of inductive logic was only developed in relation to languages which are primitive by comparison with that of Kepler's theory. But this can in no way dispose of the question of whether Kepler could have employed this logic and, if it had been employed, what

---

d. Hübner seems tacitly to distinguish between inductive logic in general and Carnap's formal theory or system of inductive logic by referring to the latter as *Induktionslogik*. I have translated this variously as "theory of inductive logic," "system of inductive logic," or occasionally simply "inductive logic" (as in the title for this section).

good it would have done him. Carnap expressly viewed the simplification and idealization that he utilized as an unavoidable evil inherent in making a beginning. Later on, one could gradually construct increasingly complex systems of logical induction, and thus find degrees of confirmation for physical theories and hypotheses. The initial primitive construction might then be more or less viewed like a rough "approximation" of the more complicated ones.[25] Hence, though it might prove difficult in practice, such an application is in no way theoretically impossible.

124

With respect to this we should probably also mention Carnap's notion that "all scientists use (roughly) the same inductive method, and that this method is close to that based on my function c*'" (by which is meant a certain selection from the possible confirmation procedures of inductive logic—Carnap designates this with "c*'").[26]

Now, even if no formal system exists yet that would enable us to calculate the exact degree of confirmation for Kepler's hypotheses and laws, nevertheless it is easy to see that if we were to calculate their value in terms of an analogous case within the confines of the primitive relations found in the languages of inductive logic, then this value would have to be extremely minimal. We found Kepler's hypotheses to be empirically very weakly supported; we found—to express it in terms of the system of inductive logic—"predictive inferences" of the most audacious sort (inferences drawn from a few planets to all planets, from two elements of the orbit to the orbit as a whole, etc.);[27] we found that the testing procedures—the "e" of this inductive logic—contained hypotheses which for their part were contained in the hypothesis "h" that was supposed to be controlled by the testing procedures, etc. Hence, if we look at things from the standpoint of Carnap's inductive logic, using its primitive languages and relations as a model—and, as we have just shown, we have a right to do this—then we must conclude that Kepler, had he known of this logic and considered it binding, could hardly have dared to hazard the formulation of his first two laws or to have stood by them.

125

But this seems to contradict what Carnap expressly warned against with respect to an identification of statements of logical induction and methodological statements.[28] The determination of the degree of confirmation as such, in his opinion, says nothing about the question of the acceptance or rejection of hypotheses, because this question belongs to the methodology. For example, in a lottery, even though the probability of winning anything on a particular number is miniscule, it is nevertheless not irrational to bet on this number. Other reasons must then obviously enter into the determination of the degree of confirmation whenever a decision concerning a hypothesis has to be made. These reasons—as they are not a proper object of his inductive logic—show themselves in Carnap's work only through the use of such vague expressions as "practical

126

decisions,"[29] "situations of life . . . within which we observe, judge, and take on beliefs,"[30] "nonlogical factors,"[31] etc.

However, even if the system of inductive logic and the methodology are not the same, nevertheless, according to Carnap's own conception, these two should be related to one another. Indeed, Carnap himself writes, "the methodology . . . develops procedures in order to utilize the results of inductive logic for particular ends."[32] On the other hand, inductive logic does not hinder the scientist from bringing in extrascientific factors for his decisions; to the contrary, it makes this task all the easier.[33] For indeed
127 what meaning should inductive logic have if not that of aiding the scientist in procuring some guidelines for this theoretical decision making as well as for his practical applications (*Handeln*)? This "logic" cannot complacently find its justification in the fact that it has initially already delimited something like the "true" and "pure" inductive relation between two statements, e and h, whereby the question of how this might be practically employed is held to be entirely external to the relation. For, in opposition to those statements peculiar to deductive logic, the statements of inductive logic have no validity in themselves at all; what is more, the axioms of this logic have already been chosen in such a way that they can supply the researcher with theoretical and practical guidelines; that is, they have been chosen in such a way that they are adequate to the general method of the scientist.[34] Even if we agree with Carnap that the system of inductive logic and the methodology are not the same, we nevertheless have to demand along with him that the two not be too sharply differentiated and that the methodology be supported by the inductive logic. If there are cases like that of Kepler, in which the methodology no longer pays any heed to this inductive logic at all (because it disregards the results of this logic), then such cases argue against this "logic." For this reason Carnap repeatedly had to limit the scope of his cited warning. He was convinced that scientists carry on as if they were governed by the numerical values of his "degrees of confirmation" (even if such were not explicitly the case), for example, in that they are prepared to invest certain sums of money in particular research projects, experiments, etc.[35] In a similar
128 manner, it was with methodological ends in mind that Stegmüller—obviously with Carnap's approval—wrote that a statement concerning inductive probability can lead us to reasonable action (*vernünftigem Handeln*);[36] that inductive logic influences the scientist in the choice of hypotheses, though it does not exclusively determine him in this matter;[37] that it helps him to make decisions with insight rather than blindly.[38] Inductive logic tells us to what degree a hypothesis is supported by the givens—the methodology must be able to utilize this information as part of a more extensive foundational context.

Hence in the final analysis Carnap maintained the strict separation of inductive logic and the methodology, especially for the realm of purely practical decisions (as, for example, betting); in the theoretical realm, however, he raised no general objection to their being closely connected.[39] An empirically weak or poorly confirmed hypothesis appeared to him as something which was scarcely capable of gaining theoretical acceptance.

Therefore, by utilizing an inductive analysis of this kind, Kepler would most likely have ended up rejecting his theory; at the very least, he would have had to view it as highly questionable. 129

It has been objected that in Kepler's case we are dealing more with an initial, hypothetical, experimental formulation of a theory, that is, with a theory in its first stage of development, whereas the system of inductive logic can only be applied to fully developed theories, such as classical mechanics, simple optics, etc. But in fact the system of inductive logic is supposed to apply to every relation of "e" and "h" (for this logic views each relation of this kind as a logical relation); and this is then always methodologically utilizable if all available information concerning "e" has been fully developed and realized. Thus Kepler would have been justified in basing his theory on this logic. But what, then, is the meaning of a fully developed theory (*ausgereifte Theorie*)? Who knows how this development might proceed—and who knows how any particular theory accepted today might look in 100 years? Kepler, at any rate, did not understand his results to be of the sort that would pertain merely to an initial hypothetical attempt; he thought that these results were as "complete" as they could be—that is, he saw them as based upon the entire body of fully elaborated astronomical knowledge of his time. There is always a here and now within which the researcher must make his decisions. And either inductive logic must be able to come to his aid in such a situation or it will never be of any help to him at all.

5.4 The Deficient Sense of the Historical
in the Thought of Popper and Carnap

From the previous investigations we find that the general worth of Carnap's inductive logic is questionable in regard to the problem of scientific decision making and that Popper's method of falsification is not always the most expedient (*zweckmäßigste*) method for science. We see Kepler acting contrary to these thinkers' ideal of "scientific-theoretical reason." 130 It has been shown that Kepler was inspired first by a mystical notion of the Sun (*Sonnenmystik*)—in which we can recognize an axiomatic precept in the terms of the previous chapter—and second by the speculative idea that the principles on which the universe is constructed are knowable— a normative precept. He took both of these notions from Copernicus, and both arose out of the spirit of the Renaissance. This solar mysticism and this speculative idea awoke in him the absolute belief in the heliocentric system and in the possibility of describing the "true movements" of this system in detail. If at first there were no better theories available for the attainment of this end than those of Tycho and the *hypothesis vicaria,* then these would just have to do—for the truth simply had to arise for man in some way or other. For the very same reasons the observational data also had to be viewed as absolutely reliable—and here we find a judicative precept. If the Sun was the symbol of God the Father, then it was of primary importance to determine the Earth's relation to it; and the exact structure of this relation had to be knowable as well—consequently one initially had to be permitted to accept what was already on hand and to use this for the attempt to determine this structure. Extrapolations from two values to all other values were justified by a belief that Divine Grace would come to one's aid in the search for knowledge about the universe; unsound mathematical analogies were sanctioned; circular confirmation arguments and those utilizing falsified theories were allowed. In the end, then, all the demands placed on the reader by Kepler are acceptable only if we allow him his mystical and speculative premises, his various precepts, if we share in his a priori decision in favor of the heliocentric system.

131   Thus Kepler's movement could be likened to that of a sleepwalker who is led on by his faith and who, despite ample distractions, is not easily diverted from his path. But once Kepler finally arrived at his goal and had developed an entirely new conception of things, he was confronted with even greater difficulties than those he had encountered at the beginning. For if we compare the *Astronomia Nova* with Ptolemy's system, then we must ask in what way had the supposedly humanly understandable construction of the universe actually become more comprehensible to the human intellect? As we have already pointed out, the elliptical form of planetary orbits baffled both Kepler and his contemporaries.

However, be that as it may, we do see Kepler proceeding on the basis of new precepts; we see him proposing a new framework within which everything is to be arranged and viewed. But this framework is historically contingent; it is not one which is taken from nature itself.

It appears to me, then, that in whatever ways the theories of science of Popper and Carnap may differ, their common and decisive weakness lies in the fact that they proceed generally in an unhistorical manner. And so it is with most of the other contemporary proposals for a theory of science as well; these all lack an understanding of the historical foundations of scientific progress, as this relates to something which goes beyond the immediate framework of science today. To awaken this understanding requires the study of the history of science. The example of Kepler shows clearly that such a study offers a corrective for the all-too-rash acceptance of methodological postulates and generalizations, as we have explained in the previous chapter.

Hitherto Kepler, and particularly the movement linking Kepler and Newton, has been brought into play precisely as the classical example for that view expressly opposed to the one espoused here. The example of Kepler was used to demonstrate that certain generally immutable methods, just like certain adequate and relevant empirical material, placed physics in a position to progress, as it were, in a self-contained manner; and consequently the observation of nature by means of this method was thought to be sufficient in and of itself, while history, and especially cultural history (*Geistesgeschichte*), was considered to be of no importance at all for this progress. Hence it was also thought that Kepler had arrived at his laws in a purely empirical manner and that Newton's gravitational law arose by way of inductive generalization out of Kepler's work. [132]

As we have now seen, however, in the first place Kepler's laws should not be viewed as empirical facts, but rather as hypotheses having a highly problematical foundation. Second, these laws are purely kinematic[e]— mass and force do not appear in them—and therefore it is impossible to infer from them by means of inductive generalization to a general dynamic law, such as the law of gravity. Third, and most important of all, when taken in a strict sense, Kepler's laws contradict Newtonian mechanics, because, according to the latter, masses exert a reciprocal attraction and thus revolve around the gravitational center of the entire system, which does not correspond with the center of the Sun, whereas according to Kepler, the Sun was fixed at the focal point of the elliptical orbit. Hence Newton altered the alleged empirical fact from which his theory supposedly grew by viewing this in terms of a new dynamic interpretation, the spontaneity of which points to an altered philosophical foundation and to historically transformed precepts taken in the sense of the previous chapter.

Therefore Kepler and the movement linking Kepler with Newton, contrary to the popular view, do not point to a nonhistorical theory of science, but rather can actually serve to support a historically oriented theory of science.[40] [133]

In conclusion we could then state the following: A theory of science without a history of science is empty, while a history of science without a theory of science is blind.[41,f] The previous example should now have made this quite clear.

e. Kinematic (*kinematisch*) refers to the study of motion independent of mass-force relations (kinematics). As such it was the basis of the physics antecedent to that of Newtonian mechanics (dynamics).

f. This is of course a reworking of Kant's famous statement: "Thoughts without content are empty, intuitions without concepts are blind." (*Critique of Pure Reason*, B 75.)

6

A FURTHER EXAMPLE: THE CULTURAL-HISTORICAL (*GEISTESGESCHICHTLICHEN*)[a] FOUNDATIONS OF QUANTUM MECHANICS

134 In 1935 Einstein, in conjunction with Podolsky and Rosen, published an article which has since taken on almost classical significance, in which he sought to demonstrate that quantum mechanics is not complete.[1] In Einstein's opinion a theory is complete when "every element of the physical reality has a counterpart in the physical theory."[2] But then what does "physical reality" mean? Einstein writes: "If, without in any way disturbing a system, we can predict with certainty (i.e., with probability equal to unity) the value of a physical quantity, then there exists an element of physical reality corresponding to this physical quantity."[3]

Einstein takes two systems, S and S', which were formerly in interaction but now are separate. Quantum mechanics describes this state by means of a Ψ-function which enables us to predict with certainty the value α' of the quantity a in S' when we have measured the value α of the quantity a in S. Since it is possible to make this prediction without introducing a disturbance in S' by measuring in S (as S and S' are separate), α' is

135 something physically real according to the Einsteinian definition, thus something which exists independent of this measurement and prior to it. Naturally, nothing in this would be altered if we had measured the value β of the quantity b in S, since then β' would also have existed prior to the measurement and simultaneously with α' as well. Now, assuming that the operators corresponding to quantities a and b are noncommuting,[4]

---

a. The German word *Geist*, along with its various adverbial and adjectival forms, is notoriously difficult to translate into English. The most direct translation is "spirit"; but this term can only be employed if an extremely close connection is maintained to the rational capability of man. As such it is best viewed as a translation for the entire complex of the Latin terms *Mens, Animus, Spiritus*. However, in common German usage, especially in compound forms, it often refers almost exclusively to mental or rational capacity or capability. In the case of the compound form in question here (*geistesgeschichtlich*), we are faced with a particularly difficult form. Hübner is referring quite literally to the history of spirit, where spirit is to be understood as the broad complex of rational and religious (spiritual) attitudes and beliefs definitive of Western culture (especially since the Renaissance). In addition, since the time of the great German historical philosophers of the eighteenth and nineteenth centuries, most notably Herder and Hegel, the term *Geist* used in conjunction with history has a marked *cultural* significance. Accordingly, *Geistesgeschichte* is at once intellectual history, religious history, and cultural history. Here I have chosen the last, cultural history, simply because it seems the broadest rendering of term.

then the wave function can determine at a given time only one of the two operators.[5] But α' and β' exist simultaneously according to the presuppositions laid out by Einstein, Podolsky, and Rosen. Consequently, the description of reality by means of quantum mechanics cannot be complete.

In response to this, Bohr admitted that Einstein and his colleagues would be correct if all disturbances were necessarily of a mechanical nature—but this is precisely what is in question. According to Bohr, there are other kinds of interference or disturbance. Consequently he draws different conclusions from the example posed by Einstein, Podolsky, and Rosen. Bohr writes:

> From our point of view we now see that the wording of the above-mentioned criterion of physical reality proposed by Einstein, Podolsky and Rosen contains an ambiguity as regards the meaning of the expression "without in any way disturbing a system." Of course there is in a case like that just considered no question of a mechanical disturbance of the system under investigation during the last critical stage of the measuring procedure. But even at this stage there is essentially the question of *an influence on the very conditions which define the possible types of predictions regarding the future behavior of the system.* Since these conditions constitute an inherent element of the description of any phenomenon to which the term "physical reality" can be properly attached, we see that the argumentation of the mentioned authors does not justify their conclusion that quantum-mechanical description is essentially incomplete. On the contrary this description, as appears from the preceding discussion, may be characterized as a rational utilization of all possibilities of unambiguous interpretation of measurements, compatible with the finite and uncontrollable interaction between the objects and the measuring instruments in the field of quantum theory. In fact, it is only the mutual exclusion of any two experimental procedures, permitting the unambiguous definition of complementary physical quantities, which provides room for new physical laws, the coexistence of which might at first sight appear irreconcilable with the basic principles of science.[6]

136

Thus Bohr disagrees with the adequacy of Einstein's criterion for physical reality in that he (Bohr) looks upon the conditions of a measurement as constitutive elements of physical phenomena. In Bohr's opinion these conditions are absolutely necessary for the unambiguous definition of physical quantities. Now, it is not possible to definitively measure the position of a particle while measuring the momentum of that particle and vice versa. Subsequently, the values of physical quantities in S' which we can predict in Einstein's example are dependent upon the measurements in S not for mechanical reasons, but rather because certain conditions have been set up which first make possible the determination of such values in any way at all. Bohr calls to mind the relativity theory, which seems to him to be built upon similar considerations. He writes:

137

Before concluding I should still like to emphasize the bearing of the great lesson derived from [the] general relativity theory upon the question of physical reality in the field of quantum theory. . . . Especially, the singular position of measuring instruments in the account of quantum phenomena, just discussed, appears closely analogous to the well-known necessity in [the] relativity theory of upholding an ordinary description of all measuring processes. . . . The dependence on the reference system, in [the] relativity theory, of all readings of scales and clocks may even be compared with the essentially uncontrollable exchange of momentum or energy between the objects of measurements and all instruments defining the space-time system of reference. . . . In fact this new feature of natural philosophy means a radical revision of our attitude as regards physical reality, which may be paralleled with the fundamental modification of all ideas regarding the absolute character of physical phenomena, brought about by the general theory of relativity.[7]

Hence, for Bohr, the arrangements and procedures (*Vorrichtungen*)[b] having to do with the measurement instruments play the role of reference systems in quantum mechanics, a role which is more or less similar to that played by coordinate systems in the relativity theory. Therefore quantum mechanics is not incomplete if it does not allow statements concerning quantities that are not definable simply because the necessary reference systems cannot be given for these quantities.

138

From these fundamental considerations we can deduce all of Bohr's other well-known categories, namely, the categories of "phenomenon," "wholeness," "individuality," and "complementarity." Under "phenomenon" he understands the irreducible "wholeness" which results from the reciprocal interaction of the measurement instrument and the object measured. He calls this "wholeness" an "individuality" because it is determined by means of the particular conditions of the measurement arrangements and procedures (*Meßvorrichtungen*), which are a constitutive part of the phenomenon. And by "complementarity" he means a connection between phenomena, which are defined by mutually exclusive measurement apparatuses.[8]

6.1 The Disagreement between Einstein and Bohr as a Disagreement over Philosophical Axioms

It thus appears that two different and opposed philosophical axioms are at the bottom of the debate as it has been delimited thus far. Perhaps it would be more exact to call these "principles" (instead of "axioms"), since we are not concerned here with statements that are directly involved

b. The term *Vorrichtung* means both a device (apparatus) and an arrangement (pertaining to apparatuses). In addition it carries a hidden reference to an intentional act (*richten*). Thus the "arrangement" of measurement instruments includes a preliminary "orientation" or view of the instruments. And all of this might be conceived as constitutive of what generally goes under the name of the measurement "procedure."

in the theory as constitutive elements, like, for instance, the Schrödinger equations. We also speak of the causal *principle* in a similar way and differentiate it from particular laws formulated in the realm of physics. Hence a principle expresses a rule of the most general significance, which 139 is then employed in a particular field, and thereby comes to serve as the foundation of particular laws. But in light of the connection with the third category presented in chapter 4 (axiomatic precepts), it seems more useful to me in what follows to speak of principles as "axioms." It would serve no purpose with respect to the present problem to make an overly pedantic distinction here; at most this could only complicate matters needlessly and impede clear understanding. Therefore, where the term "principle" has been expressly adopted by linguistic convention, we will keep it so as not to inconvenience the reader too greatly—as for instance in the case of the "causal *principle*" or that of the "cosmological *principle*" (cf. chapter 10).

I will now try to formulate the philosophical axioms lying at the root of the debate in question in a more generalized form and one which is removed from the particular case under consideration. According to one of these axioms, that underlying Einstein's position, reality consists of substances that have properties which remain unaffected by the relations between substances. According to the other axiom, that of Bohr, reality is essentially a relation between substances and measurement uncovers an intrinsic state (*einen Zustand an sich selbst*): for Bohr a measurement constitutes a reality. For Einstein relations are defined on the basis of substances; for Bohr substances are defined on the basis of relations. These general philosophical positions make up the basis of the discussion. I will signify the first axiom by the letter "S" (since it holds substances as basic), the second with the letter "R" (since it holds relations as basic).

Neither Einstein nor Bohr was, however, in a position to prove the 140 validity of the axiom characteristic of his respective view on the basis of the examples they studied, or to refute that of the other. Let us recall that Einstein gave his criterion for reality in the form of an "if-then" proposition: "If, without in any way disturbing a system, we can predict with certainty the value of a physical quantity, then there exists an element of physical reality . . . etc." If A, then B. Now, according to Einstein, in the case of two separate systems, S and S', A is true since S' is not disturbed. Hence B is also true; and there really do exist quantities in S', and indeed these exist independent of measurements in S. However, when Einstein says that A is true, he has already accepted axiom S, since he is already convinced that a system which has not been mechanically disturbed has not been disturbed at all. Moreover, by presupposing this principle, he concludes that such a system possesses intrinsic properties.

Accordingly, the example chosen by Einstein, Podolsky, and Rosen does not establish the validity of axiom S; rather it is only an interpretation of this axiom on their part. Consequently they are not then in a position to refute Bohr's view. On the other side, however, Bohr is in the same situation. For him, A is not true since he believes in axiom R. Therefore he has not refuted Einstein and his supporters; rather he has only shown how it is possible to respond to them, how it is possible to interpret the case in question so that the completeness of quantum mechanics can be defended.[9]

6.2 Is Bohr's Philosophy Idealism?

141 Thus it was axiom against axiom, and it was necessary to seek new arguments in order to push the discussion further.

In a further development of the debate, we encounter considerations of a purely philosophical nature, as is shown in what follows. I quote Einstein again: "What I dislike in this kind of argumentation," referring to Bohr and his supporters, "is the basic positivistic attitude, which from my point of view is untenable, and which seems to me to come to the same thing as Berkeley's principle, *'esse est percipi.'* "[10] Blokhintsev expresses a similar view: "Thus we see," he declares, "that all problems of quantum theory, according to N. Bohr, are to be viewed as problems

142 pertaining to the interaction of the instrument and the micro-object, as problems—and with this he abandons the firm footing of physics—of the interaction of subject and object. The fundamental methodological mistake of the theory of complementarity resides in this fact as well: In light of this conception, the laws of quantum mechanics lose their objective character and become, as it were, laws resulting from the manner and means in which man perceives the appearance of the micro-world. And this is idealism."[11]

However, this sort of critique of Bohr goes amiss in that axiom R is distinct from Berkeley's idealistic "esse est percipi," even though it must be admitted that Bohr himself did not always recognize this clearly. The relation between an instrument and an object is not at all the same as that between a subject and an object in the sense of idealism. The first relation is purely physical in nature: it is the relation between objects, even if it is also the condition of the possibility for such perceptions. If we were to press down on the accelerator of a car, we would not think of this as something subjective, even though it is also the expression of a certain desire, namely, the desire to drive. Of decisive importance here is the fact that we can essentially replace the subject with an object in such a case (for example, replace the driver with an automatic device). Above all, then, we must always remember that axiom R deals with relations between objects in general, among which the measurement process is only a special instance. Hence this axiom does not actually correspond in a precise sense, as it is often taken to, with the expression "Being is

being-measured"—something which we had already referred to in chapter 2.

Strictly speaking, this formulation is even misleading, because it suggests that the subject cannot be excluded from the constitution of the relation if we accept axiom R. But when, for example, the conditions defining the position of a particle do not exist, then the particle has no position at all—just as the legendary land of Atlantis has no location. And when the conditions defining the momentum of a particle do exist, then it does in fact have a momentum—just as Berlin has a relative position on the surface of the Earth. It is irrelevant whether an observer has produced these conditions himself or whether he found them already established.

Accordingly, there are no necessary connections between Bohr's philosophy and positivism (or idealism), as has so often been asserted.[12] This does not mean, as I have already said, that Bohr was always clear on this point. Axiom R, in my opinion the core of his philosophy, is in itself neutral with respect to different standpoints in terms of the classical theory of knowledge, since it does not contain a direct reference to the subject and no statements about the ego can be directly deduced from it. But even if the attempts of Einstein, Blokhintsev, and others to reproach Bohr's theory as being positivistic or idealistic fail, nonetheless, in light of the course of the discussion as we have depicted it here, Bohr's axiom does seem to be more of a stipulation than a well-grounded statement. Hence, as we said before, this second phase of the debate does indeed hinge on purely philosophical arguments.

6.3 The Example of the Cat
In the same year that Einstein, Podolsky, and Rosen published their ar-
ticle, Schrödinger wrote his well-known essay concerning "Die gegenwärtige Situation in der Quantenmechanik" ("The Present Situation in Quantum Mechanics"), in which he also considers an example of particular importance for the matter under discussion here.[13]

He presents us with the picture of a steel vault containing a cat and a radioactive substance. Then he assumes that the probability for the decay of an atom of this substance in the course of an hour is just as great as the probability that it will not decay in the same amount of time. In the vault there is also a device which gives off hydrocyanic acid when such radioactive decay occurs, so that the cat will be killed. Now according to quantum mechanics the state of the atoms is not well defined, and consequently neither is the state of the cat. According to axiom R this means that the cat is neither really dead nor really alive.

For Schrödinger, as well as for Einstein, who formulated a quite similar example,[14] this is completely absurd. The cat is a macroscopic object and is in a well-defined state, being either alive or dead. Consequently, the state of the atoms, which in this case strictly determines the state of the cat, must also be considered just as well defined.

Once again quantum mechanics appears to be incomplete. But again this kind of argumentation is not conclusive.

145 The state of the cat can be viewed as undefined insofar as it is dependent on the state of the atoms in the radioactive substance. Let us say that when the cat is alive, the state of the atoms is A, and A' when the cat is dead. Then, in accordance with axiom R, neither A nor A' exists; and consequently the cat also has no real state as long as its state is related to the state of the atoms. But in opposition to this, the cat is actually dead or alive according to certain medical devices by means of which we measure the pulse rate, etc. In a similar fashion we can say that Berlin has no location in relation to Utopia, but it has a well-defined location in relation to Washington, D.C. According to axiom R there are no intrinsic states, rather only states relative to something. Consequently, the argumentation of Schrödinger and Einstein rests on an equivocation. They argue as follows:

a) State X is well defined.
b) In quantum mechanics it is not well defined.
c) Therefore quantum mechanics is not complete.

But they overlook the fact that in Schrödinger's example in (a) X means the state of the cat relative to certain medical apparatuses, whereas in (b) X means its state relative to the radioactive substance.

In Bohr's language, in (a) we mean the "wholeness," "cat and medical apparatuses," whereas in (b) we mean the "wholeness," "cat and radioactive substance." Hence we are not dealing with the same X in (a) and (b), which would be simultaneously both well defined and not well defined. Consequently, the conclusion of Schrödinger and Einstein is not correct. However, it must be remembered that this is the case only if axiom R has

146 been presupposed. For those who believe in axiom S, the cat is either dead or alive and its relation to other objects (instruments, substances) has no meaning at all. So in the end the example of the cat poses no real problems for either the supporters of quantum mechanics or its critics; rather the axiom in question is just as easily interpretable by one side as the other.

6.4 Operators for Unmeasurable Quantities in Quantum Mechanics
Up till now I have discussed a few of the most important attempts to show that the interpretation of quantum mechanics by Bohr and his disciples leads to untenable results. But can axiom R be brought into any kind of complete agreement with the formalism of quantum mechanics at all?

In 1952 Wigner demonstrated in his article "Die Messung quanten-mechanischer Operatoren" ("The Measurement of Quantum-mechanical Operators") that the greater part of the possible operators in quantum mechanics do not represent measurable quantities.[15] This means that there is no possible reference system (measurement apparatuses) for these quantities and that, subsequently, according to axiom R, they have no reality, even though they are exactly defined through the formalism of quantum mechanics.

If we now maintain, as in axiom S, that properties of physical entities are independent of measurements, since properties belong to these entities in this view irrespective of their relations with other entities, then from the standpoint of axiom S such measurements have only a secondary (*zweitrangige*) meaning and can never be put forward in any rigorous 147 sense. Thus the formalism of quantum mechanics does not appear to rule out axiom S completely. To the contrary, in many respects this formalism seems to agree even better with axiom S than axiom R, since, according to Wigner, it permits the introduction of quantities that must be viewed as intrinsically existent. But the price paid for this advantage is high, since it entails the contradiction of another generally accepted axiom: never to admit of quantities which cannot be measured at all. I do not believe that Einstein thought it possible for there to be a contradiction between axiom S and one of the fundamental ideas of his relativity theory—namely, that every definition of a physical quantity can be conceived operationally, and thus with respect to the instruments of measurement. But be that as it may, in this phase of the discussion we can find advantages and disadvantages on both sides of the issue. So once again we must affirm that the real fight is one over axioms.

6.5 Quantum Logic, Interphenomena, von Neumann's Proof, and Indeterminism

There were those who apparently believed that they could put an end to the debate once and for all by using a special, supplementary logic, often called quantum logic. Reichenbach, for instance, attempted a formal analysis of the Einstein-Podolsky-Rosen paradox by means of this logic. He summarized Bohr's standpoint in the following statement: "The value of a quantity prior to a measurement is different from the result of the measurement."[16] Let the letter A stand for this proposition. Then ac- 148 cording to Reichenbach's view, as regards Einstein's initial example, A cannot in fact be true, at least not with respect to system S', since it is separate from system S, within which the measurement takes place. So Einstein would be right on this score. But, on the other hand, he would be wrong in concluding from this that A has to be false, since, according to quantum logic, it might also be indeterminate. Consequently, even if A is not true, we cannot therefore conclude from this that the proposition "the value of the quantity after a measurement is the same as before," which expresses Einstein's view, is true. So for Reichenbach the argumentation of Einstein, Podolsky, and Rosen collapses; but neither does he support Bohr, since Reichenbach himself does not in fact think that proposition A is true.

Moreover, we should not confuse quantum logic with formal logic. Quantum logic (as I will try to show more fully in the following chapter) is nothing but a special calculus, suited to a special interpretation of quantum mechanics and a special formulation of quantum-mechanical laws. Hence it is of little help to prove something by means of this logic, since it is just as questionable as this interpretation and formulation. Quantum logic cannot be granted the same weight as formal logic, which, as Leibniz said, is valid in all possible worlds.

But even so, we might still discuss Reichenbach's philosophy of quantum mechanics somewhat further. Like Einstein, Bohr, and Schrödinger, Reichenbach also employs a particular example, i.e. Young's famous experiment (which will not be described here). Reichenbach points out that when we interpret this experiment with the help of certain assumptions pertaining to the existence of well-defined entities that have no relation at all to the measurement process, and thus are termed "interphenomena," then along with this we must accept certain causal anomalies or redundant precepts that can never be verified, falsified, or used for predictions. Such entities might be things like "corpuscles" with position and momentum or "waves" radiating in space.[17] By causal anomalies he means the break with the principle of action by contact (*Nahwirkungsprinzip*), whereas by redundant precepts he means the values of the position and momentum (of such entities) *between* measurements—values which can never be determined by measurement.

Reichenbach recognizes with perfect clarity that neither the principle of action by contact nor the ban on redundant precepts is a sacred cow and that here we are concerned with axioms. But he does not attempt a more detailed discussion of these axioms; it is for this reason that his efforts remain unsatisfactory. Further, he only studied special kinds of hidden variables, either particles or waves. And in the intervening years several theories have obviously been developed which seek to avoid the difficulties arising from the consideration of such particulars. Examples would be the theories of Bohm and Bub.

But even if it is impossible to prove by means of a special quantum logic that no theory of this type can be true, might it perhaps nevertheless be the case that von Neumann's famous proof establishes this fact?

Briefly stated, this proof runs as follows:[18] By a "pure case" ("*reinen Fall*") we mean a collective grouping (*Gesamtheit*) of N systems, all of which have the same state-function, or otherwise expressed, all of which have the same probability distribution (expectation value) for physical

149

150

quantities. Now, if hidden parameters—intrinsic quantities—do exist, then it would also have to be possible to reduce the probability distribution of a pure case to the distribution of real states. We would then, in fact, have a mixture, that is, a collective grouping composed of subgroupings, each one of which is again a pure case. But von Neumann proves that this reduction is impossible, because the predictions based on a pure case differ from those based on a mixture.[19] He also points out that such a reduction would imply the possibility of representing the collective grouping of systems of a pure case as one subdivided into dispersion-free subgroupings, where every element of these subgroupings has the same value u_k for the quantity U. But dispersion-free groupings cannot exist, because if they did, then the trace of the density matrix for a pure case would not equal 1; and this would contradict its definition in Hilbert space.[20]

151

Now, von Neumann's proof can only have a limited meaning because it presupposes quantum mechanics, which, being an empirical theory, obviously cannot be true by necessity. At best this proof might be thought to show that every type of theory implying hidden parameters will be incompatible with quantum mechanics. But in fact this is precisely what it cannot prove. For what has von Neumann actually proved? He has demonstrated that the formalism of quantum mechanics does not permit any hidden parameters such as those which are defined within the limits of this formalism and which partly overlap classical quantities. Therefore the concept "hidden parameter" is used in such a limited sense by von Neumann that the proof cannot have any universal meaning for all hidden parameters. Bohm and Bub have, for example, introduced special kinds of hidden parameters, such as nonclassical potentials or quantities that are determinable within extremely short intervals of time following the measurement, but which dissipate again immediately thereafter.[21] It follows from this that von Neumann's axiom, "Av(R) + Av(S) = Av(R + S)"—where R and S are observables—is no longer universally valid. Therefore (for Bohm and Bub) quantum mechanics is to be viewed only as a limited case, that is, as a statistical theory that can be deduced from a deterministic theory in which the quantities are qualitatively different from those of quantum mechanics. Indeed, theories like those of Bohm and Bub have their own special problems; but what is essential in the present context is that von Neumann's proof is not the means by which these theories, and the parameters associated with them, can be refuted.

152

153

Here it is especially interesting to note that particularly in the case of the theory of hidden parameters, like the one proposed by Bub, axiom R is expressly used as a foundation. Bub writes: "The deep intention behind the development of hidden variable theories is the realisation of a 'natural philosophy' which incorporates a concept of 'wholeness' as a new ontological basis."[22] It seems to me that by "wholeness" Bub means Bohr's notion of axiom R. He is obviously thinking of this principle when he says that it is the essentially revolutionary and progressive element of the new physics; and the only problem resides in the fact that Bohr has not developed it with sufficient consistency. Thus we can now state with certainty that axiom R is manifestly just as compatible with determinism as axiom S. This is of great importance.

I do not think that either Einstein or Bohr actually realized this. They disputed matters in terms of axiom S and axiom R; but in reality they both appear to have been animated at a deeper level by the question of determinism versus indeterminism. Einstein's famous remark "God doesn't throw dice" points clearly to this. Accordingly, the intellectual struggle depicted here is at least just as much a struggle over the philosophical categories of "reality" and "substance" as one over the category of "causality."

6.6 How Can the A Priori Axioms Underlying Quantum Mechanics Be Justified?

Looking back, we can now state that, while, on the one hand, philosophical axioms do more or less constitute the basis of the debate about the nature of reality in quantum mechanics, on the other, these axioms were not sufficiently discussed, but rather merely taken as self-evident truths; not to mention the fact that nothing was said concerning whether such axioms could actually be justified or refuted. Accordingly, I will now turn to the question of how such justifications are possible in any sense at all. Attempts have been made essentially in three ways:

1. by means of purely philosophical considerations;
2. with the aid of experience;
3. on purely methodological grounds.

I will discuss these three possibilities in the order given.

In terms of the cursory overview and synopsis attempted here, only a very few of the purely philosophical considerations of physicists could be treated. But it is obvious that almost all of the important physicists who have been concerned with questions pertaining to the foundations of quantum mechanics have philosophized in a more or less close connection with the axioms which have been discussed. The reflections of these physicists are at least partially based upon extensive philosophical studies (and this has proved to be the case in the most recent historical research as well, for example, that of Jammer and Meyer-Abich).

Beginning with Einstein by way of example, we then see that he was deeply rooted in the Cartesian tradition, and hence was influenced by ideas about the divine construction of the universe ("God does not throw dice'') that can be specifically referred back to Galileo and Kepler. The Cartesian tradition is the source of the doctrine that physical reality is composed of well-defined substances that enter into relations (interaction) with one another. These substances are well defined in the sense that they have mass and a degree of velocity; they interact in that they alter these "primary" velocities by means of forces which they exert on one another "secondarily." What pertains essentially to the substances and what gets altered with respect to them through external agency can thus be fundamentally and clearly distinguished. The "Cartesian tradition" should be understood here as a fundamental ontological conception about reality and should not be confused with the Cartesian philosophy as such (cf. chapter 9). Nevertheless, this fundamental ontological conception does, in a certain sense, receive its initial prefiguration with Descartes, a prefiguration that would later be refined but in no way essentially altered. Hence, even if Newton might also be called the true father of classical 156 physics, it is nonetheless the case that with respect to this conception, despite important modifications, Newton built upon the foundations laid by Descartes.

Moreover, we can then see that Bohr was wrong when he reproached Einstein by asserting that, contrary to his (Einstein's) stated position, his relativity theory was itself representative of a conception which—to express it in the terms used here—has its origin in axiom R, though we would admit that Einstein's view does bear certain similarities to Bohr's standpoint. Einstein's theory does indeed maintain the relativity of all phenomena to reference systems. But this relativity exists properly only at an ontologically quasi-lower level, that is, on that level where the reference systems (Earth, Sun, etc.) are regarded as the true reality. It was Cassirer who then removed the relativity theory from this last vestige of "geostasis" ("*Erdenrest*") by stating that this theory attains to a level which is of a quasi-higher ontological status, hence to a unity of the description of nature which is neutral for all systems of reference. Regardless of the standpoint from which we view things, in the general field equations we find that the relativity and "subjectivity" disappear once again; the states are described as covariant with respect to all reference systems, thus independent of and detached from conditions pertaining to the possibility of their being experienced. But in this way physics once again comes into line with the ontology originating in the Cartesian tradition, even though what is now understood by "well-defined substances" is no longer the same as before (since now the functors "mass" and "momentum" are differently defined).

Einstein's deep belief in the determination of nature is without doubt stamped with that kind of spirituality (*Religiosität*) which arose in the course of the Renaissance and remains deeply embedded in Western con- 157 sciousness up to the present day. Here we are concerned with the belief already discussed in chapters 4 and 5: that God purposively made the world with a view toward rationality and that accordingly the "book of nature" is written in the language of mathematics. Thus it is not divine caprice, not irrational accident, that dictates nature, but rather logical necessity and harmonious justice. For this reason the equivalence of reference systems appeared to Einstein as an expression of the harmony of the universe. (Chapter 10 will have more to say about this.)

With regard to Bohr, his thought reveals his lifelong interest in the philosophy of Kierkegaard and James, as well as the fact that he felt himself inspired by the Danish poet and prose writer Møller. There obviously exists a certain analogy between Kierkegaard's dialectic and Bohr's principle of complementarity; at any rate, Bohr expressly took this to be the case. What is most important here is that Bohr built upon Kierkegaard's view of the relation between subject and object—a view which initially results from the analysis of the subject carried out by the subject. Since the subject is an entity which reflects upon itself, it is itself object. But this is only one side of the coin, so to speak, for it is not merely object, but rather subject-object. We can never consider the two together with equal clarity, and we can never have the one without the other. When the subject becomes object to itself, its subjectivity disappears behind its objectivity. But precisely for this reason, it only grasps itself one-sidedly in the objectification and must accordingly negate its mere objectiveness once again. Thus it is thrown back anew on its subjectivity, which shuns this objectified state, only to become once again newly objectified, etc. It was precisely this description of existence Bohr found in Møller's story *The Adventures of a Danish Student,* where the main character constantly sought in vain to think himself. He thinks himself as a thinking; but then he becomes aware that he is a thinker, who thinks himself as a thinking, etc. This transition from subjectivity to objectivity and vice versa is not temporally conceivable for Kierkegaard, for then it would again be an objective experience. The transition takes place in an "instant," which he designates as a "leap"; and he views this leap as an act of choice. Moreover, this kind of dialectic does not remain limited merely to the reflective ego; it is carried over, as it were, to the general relation between subject and object, and hence to the concept of truth itself.

These thoughts, transmitted in particular by the Kierkegaardian thinker and friend of Bohr's father Høffding, could obviously be found in other philosophers as well, especially since here the specific dialectic of temporality and eternity characteristic of Kierkegaard's thought has been left out and treated only in abstract form. But in the present context we really only need to point out how Bohr was influenced by Kierkegaard and not what Kierkegaard actually meant. Viewed in this way, it is not so surprising that Bohr could have felt himself equally drawn to James, who is worlds apart from Kierkegaard. Thus Bohr seems to have been

158

fascinated by the same aspect in James's thought as had interested him in Kierkegaard's, namely, the analysis of consciousness. James laid this out most precisely in his *Principles of Psychology*. There we again encounter the question of how one's own thinking might possibly become objectified. In order to deal with this, James had recourse to the decidedly important dialectic of "substantive parts" and "transitive parts." The substantive parts relate to what is immediately apparent, propositions, words; but underlying this level of the immediately conceivable are, so to speak, the transitive parts, relating to the characteristic flow of thought, the transitions. If we wish to conceptualize these, we must change them into substantive parts, thereby destroying them; and inversely, if we concentrate on the substantive parts, the transitions once again slip away. So here again a kind of complementarity seems to be the rule. Finally I might also mention James's doctrine that, viewed in a strict sense, consciousness never knows anything fixed. Everything is known only conditionally, and the conditions change; no object every offers itself to consciousness in a manner which would allow it to be removed from such relations.[23]

In light of all this, one might well ask why it is that philosophical considerations do not play a decisive role in the discussions of quantum mechanics. Why are they relegated more or less to a level of peripheral importance? Why are not the axioms which have been mentioned here always dealt with first of all in a purely philosophical manner, distinct from particular problems of physics?

The answer is simple: Today most physicists look upon philosophy as something which is useful for spurring on thought and as an interesting supplement; but they do not earnestly think that it is capable of offering rigorous proofs. Being children of an age conditioned by a more or less positivistic mentality (*eines positivistischen Zeitgeistes*), they have—if I might be allowed to express it thusly—some "philosophical complexes"; thus they believe above all else in experience. In this respect they are fundamentally different from the physicists of the great classical period, Galileo, Kepler, Newton, and their disciples. Here I cannot go into the details of the cultural (*geistige*) development that has resulted in the antimetaphysical or, more simply, antiphilosophical stance. However, it might at least be said that this stance is essentially rooted in the conviction that there is no absolute evidence and no pure reason. Hence all attempts to answer questions about reality, causality, substance, etc., must, in their opinion, go awry in that they are removed from determinant physical conceptions. Lack of faith in philosophy is thus also lack of faith in absolute conceptions and eternally valid insights. In fact, if philosophical axioms are to be justified through abstract philosophical thinking alone, then how can we avoid deducing these axioms from others which, in the end, are to be viewed as self-evident, absolutely valid, or necessary in some way or other? And how can we in all seriousness allege that these axioms do in fact have such characteristics? One need only glance around to see that whether it is a matter of Descartes's faith in Euclidean geometry or in his particular formulation of causality, or whether we are concerned with Kant's belief in transcendental apperception or Hegel's belief in the necessity of the beginning point in his logic, etc., everywhere we find axioms taken to be first principles (*letzte Axiome*), which are held to be valid by reason of pure insight, pure reason, or pure thinking. But over and over again history teaches us that all axioms which have been viewed as eternally valid have been given up at some other time. Moreover, things that are held to be almost trivial at one time can, in the context of some later view, be pointed to as the result of lengthy debates and investigations, where one has come to trust in rather complicated ideas and become accustomed to these. The history of science, running from Aristotle to the present, is essentially a history of axioms and the revolutionary overthrow of these axioms. It seems to be an inextirpable characteristic of mankind that every new revolution is looked upon as the revelation of an absolute truth, or at least of something approximating such a truth, which then merely stands in need of improvement. But even if we are at least able to understand to a certain extent the widespread skepticism among physicists concerning the attempts to justify, by means of philosophical considerations, the axioms which they must employ, nevertheless, as shown in the previous chapters, this in no way gives us the right to share in their belief that experience can supply the foundations for which they are looking.

It became apparent, especially in chapter 3, that proofs based on experiments and successful theories do not constitute, in any respect, absolute high courts against which objections can no longer be raised. Einstein, de Broglie, Bohm, and Bub, etc., did not capitulate in the face of the great success of quantum mechanics. Their general stance was theoretically justified on the basis of the fact that from a scientific point of view there are no absolute facts, but rather only relative ones, that is, relative to particular presuppositions and a priori precepts. But what is the meaning of not capitulating in such a situation? It means that one holds on to certain axioms while not accepting others. This is rather confusing. On the one hand physicists more or less try to avoid philosophy in that they base their research on experimental and empirical foundations; on the other hand they act, at least implicitly and perhaps without full consciousness that they do so, as if they too have doubts about experience, thus holding fast to certain axioms in an a priori manner which they have decided in favor of. If this unwillingness to surrender axioms is not to be termed dogmatism, then they must at least look for an explanation of it. But where might they look for this explanation if not in philosophy? And if this is the case, can we not then liken their situation to that of Ulysses caught between Scylla and Charybdis? They can believe in neither pure reason nor pure experience, since neither the one nor the other really exists.

Some thinkers believe that there is a way out of this dilemma if we attempt to justify our axioms on purely methodological grounds. Bub is a good example of this with his theory of hidden parameters. In agreement with other thinkers—for instance, Feyerabend—he holds it to be inadequate to call a theory into question only after it has been shown that it no longer yields new discoveries.[24] To the contrary, according to Bub, we should already be developing alternative theories when a theory is still actively producing results, since it is only in this way that we can look for something that is not interpretable within the old framework, but is only intelligible within a new framework, and thus is something really new. History also teaches us that an old thesis is never abandoned until the problems with it have been made apparent by means of a counter-theory. Hence, waiting for new discoveries and the consequent rejection of an established theory would only lead slowly to sterility and dogmatism. Accordingly, Bub defends what he calls his "ontological principle" expressly by means of reference to the fact that a countertheory is founded on it which is opposed to widely accepted principles and theories. Obviously, he also seeks to establish this through the use of experimental proofs and empirical results; but his "principles," which I would rather call axioms for the reasons already stated, have already been justified in a certain sense by the methodological arguments which were just presented. Hence we see, for example, that Bub presents us with something like axiom R, along with other axioms, precisely because this seems to him to be a good strategy for new research and for extending the possibilities of scientific progress.

But this form of argumentation and justification can only be discussed if we treat the following questions:

1. Is the strategy employed really a good one?
2. Should we accept the scientific goals that are to be attained by means of this strategy?

At present I do not want to deal with these questions; rather I only wish to indicate that this debate cannot take place within the framework of physics alone. To the contrary, such a discussion entails renewed philosophical considerations as to what the goals of science should be and why in certain cases holding to certain theories leads to dogmatism and sterility, whereas in others it does not; why sometimes this resolve to defend fundamental positions must be given up; why this defense cannot be founded on the view that these positions are necessarily true; what is to be understood by progress, etc., etc. Hence, if we desire to escape philosophy by turning to experience or by making use of methodological

164 considerations alone, we find that we will nevertheless always end up involved in precisely that from which we were fleeing: philosophy.

But perhaps this *horror philosophiae* which modern physicists so often feel is rooted in an incorrect interpretation of what philosophy should be. Perhaps our present-day physicists, who are so disposed to being revolutionaries in physics, are tied too closely to tradition and are too conservative when it comes to philosophy. Does philosophizing necessarily mean seeking final solutions to problems? Perhaps we should not connect philosophy with the idea of absolute knowledge, eternally valid insights, absolute reason, necessarily valid axioms, and self-evident, final revelations.

Let us then review matters one more time. We have seen that we cannot avoid the use of certain a priori axioms when we discuss the foundations of quantum mechanics, and that such a prioris, to use Kant's language, are conditions of the possibility of experience. But if, unlike Kant, we cannot accept any kind of necessary insights belonging to a pure reason, we might nevertheless still be capable of justifying a priori statements by referring them to the particular historical situations in which they arise. As I have already noted, nearly all a priori axioms have been abandoned or altered in the course of history. On the one hand this shows that it was questionable to view such axioms as absolute; but on the other hand it is nonetheless the case that changes, developments, and even revolutions do not simply fall from heaven. All of this can be rendered scientifically

165 intelligible by means of the history of science. There is a kind of historical reason and historical contingency that is neither pure necessity nor pure accident. Accordingly, if there are any scientific justifications for axioms at all, they will be historical in nature.

Thus, if we take the discussion of the foundations of quantum mechanics as an example, we will also have to discuss axioms R and S—the first with a view to a particular experimental situation in physics, and the second with a view to a certain intellectual-cultural (*geistige*) situation, which is tied to a tradition that is still alive and apparently cannot be made to conform to the situation in physics. However, to the extent that physicists discuss these axioms, they will have to philosophize in a comprehensive manner; and this they cannot do by way of a private hobby which they casually pursue. To philosophize about axioms R and S means to take into account the historical tradition encompassing Aristotle, Galileo, Kepler, Descartes, and extending right down to the present, as well as to discuss the historical background constituted in a sense by James, Kierkegaard, etc.; finally it entails the discussion centering on the history of quantum mechanics itself. Only by doing this can we avoid that view in which the various positions dogmatically appear as self-evident and in which the problems seem already to have been solved. Only historical considerations allow us to understand these different positions and the meaning of their axioms. Moreover, such considerations also have an important critical function. They show us that nothing is necessarily true, but rather that every position is dependent upon the particular conditions of its origin.

In addition I should like to offer yet one more brief and admittedly somewhat simplified observation which might serve to elucidate the historical background of axioms S and R in a manner which goes beyond the immediate sources that have already been discussed, and which formed the basis of the theories of Einstein and Bohr. The ancient skeptics made   166
reference to the thoroughgoing relationality present in the connection between things and the knowing agent as well as among the things themselves in order to show that it is impossible to grasp these entities in their true being, i.e. in their in itself. In contrast to the skeptics, we see other thinkers, for instance, Aristotle, who limit this relationality to the level of certain categories ($\pi\rho\delta\varsigma$ $\tau\iota$), which have as little significance for what the things are in their essences as substances as does the view that in relation to person Y, person X may be shorter, while in relation to person Z, X may be taller—since nothing is then stated about the essential characteristics of X. Logically this latter view is expressed in the fact that the categorial judgment having a single-place predicate is viewed as the fundamental form of all judgments, whereas predicates of more than one place are thought to designate ontologically nonessential phenomena. The results of modern physics with its radical mathematical depiction of nature, especially that of Descartes, and the central significance thereby granted to the concept of functions, have brought about several changes in this view. Substances are now determined in terms of a nexus of relations (movements and forces), but nonetheless in such a way that once again it is only in terms of constant essential properties that the changes and modifications in substances can be explained. (Hence every body has mass, location, and velocity.) But these properties as such, which describe the "essential state" of a body, are not thought to be given relationally. Axiom S, even though already hard-pressed by axiom R, can thus still be maintained at this stage of the argument. Nor is this altered by Kant's transcendental philosophy, from which he derived his dynamic meta-   167
physics of nature. For even though the object may be an appearance as well in this view, as such it nevertheless is still subsumed entirely under the conditions just pointed out: it has mass, location, and velocity. Within this historical context, the relativity theory appears as a capstone and high point, most essentially because here axiom R is granted, as it were, the greatest possible free play while still remaining tacitly constrained [by axiom S]. With respect to this, the first real change only occurred, as noted earlier, under the influence, on the one hand, of dialectical philosophy and, on the other, of the new occurrences in microphysics.

7

CRITIQUE OF THE ATTEMPTS TO
CORRELATE QUANTUM MECHANICS
WITH A NEW LOGIC

168   The results of the previous chapter stand in need of an important supplement. There we warned of the mistake involved in viewing Reichenbach's quantum logic as offering a final solution to the dispute between Einstein and Bohr, since such a position involved a renunciation of all historical connections. Now we must explain this in more detail.

Today there is still a widespread opinion that quantum mechanics has led to a new logic, and hence has at the same time revealed structures of language which had not been adequately noted before. The older logic is thus thought to possess only a limited validity over and against this new one; indeed, in certain cases pertaining to quantum mechanics it is even thought to prove false. From this, certain philosophical conclusions have been drawn: For instance, it has been asserted in recent times that the formal foundations of all thought—and obviously logic is concerned with these—have been transformed by the events of modern physics. These foundations are no longer accepted as quite so universally valid and immutable as before. Along with this it has also been asserted that certain allegedly new insights into the essence of thought have come to light as well as insights into the nature of speech in general. Thus quantum mechanics is thought to possess a kind of universal significance, a significance extending far beyond the realm of physics alone.

7.1 Von Weizsäcker's Attempt

169   Several of the works of C. F. von Weizsäcker are especially representative of such ideas. In these, classical logic is understood merely as a methodological a priori that has to be used in the formulation of quantum logic. Moreover, in terms of this modern view, it is quantum logic which is thought to be the *true* logic while classical logic is only held to be a limited case of the former. It is thought that one must create a logic that is "suited to" present-day physics; accordingly logic is only held to be true in the sense in which a physical theory is thought to be true, that is, not absolutely, true, but only true in the sense of admitting of constant improvement. "It would be conceivable," he writes, "that for us structures of being have become apparent through the examples of present-day physics which would not be compatible with the ontological hypothesis lying at the basis of classical logic."[1]

Now it remains uncertain whether there are in fact certain hypotheses of one kind or another—for instance, ontological ones—lying at the basis of classical logic. But here it is of special interest to point out that according to von Weizsäcker's conception a certain empirical development within modern physics has produced a change in logic. In this way logic is swept up into the ongoing process of change characteristic of the natural sciences. At the same time it obviously loses that inviolate a priori status that has been looked upon for ages as its exceptional characteristic. And because it is now only granted the status of a methodological a priori, useful only for the production of new forms, it also falls under the sway of the unsteady light of empirical improvability.

In light of these claims I pose the following question: Has quantum 170 mechanics really developed a new logic, and thereby cast doubt on the value of traditional logic? In order to discuss this let us begin by taking a look at the so-called Youngian dual-slit interference experiment (figure 7).

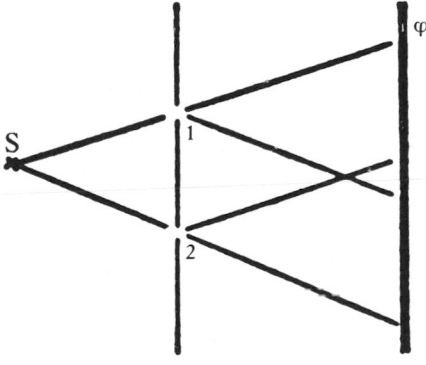

Figure 7

Here electrons from source S pass through a shield with two slits and strike a photographic plate. Experientially speaking, the position of the point at which a particle makes contact with the plate cannot be exactly predicted; rather this can only be described by means of a probability function P. Now, if only slit 1 is open, then we get P_1; if only slit 2 is open, then P_2. But if both are open, then we have P_{12}. We would then suppose the following equation to hold here:

$$(1) \qquad P_{12} = P_1 + P_2 .$$

However, the experiment shows this equation to be false. If $\Psi(x)$ is the probability amplitude introduced by quantum mechanics, then the state of affairs would be correctly formulated as follows:

(2)
$$P_1 = |\Psi_1|^2, \; P_2 = |\Psi_2|^2, \; P_{12} = |\Psi_{12}|^2,$$
$$\Psi_{12} = \Psi_1 + \Psi_2.$$

Accordingly, we must question the presuppositions by which one arrives at the false equation (1). These are:

1. Electrons are material corpuscles.
2. Every corpuscle passes through either slit 1 or slit 2. *Tertium non datur (TND)*.[a]

The proponents of the so-called quantum logic show no compunction at giving up the first of these two presuppositions. In fact, it was on the basis of this experiment that Young came to his conclusion that light has the nature of a wave. But these same proponents choose as well (for reasons that we cannot go into here) to give up the second presupposition—namely, the principle of classical logic—and believe that, as a result of this, logic must surely be modified.

First, then, let us look again at the rather transparent and easily approachable "three-valued logic" developed by Reichenbach.[2] He named this a "three-valued logic" because he added a third value, that of "undetermined," to the other two values normally applied to statements, "true" and "false." To this end, Reichenbach introduced the following schema:

SCHEMA 1

1	2	3
A	$\overline{A}$	~A
T	U	U
U	T	F
F	T	T

172
Column 1 indicates that a statement A can be true, undetermined, or false. In column 2, the negation of A, denoted as $\overline{A}$, is defined by reference to the three values of A in column 1; and thus the negation cannot, as in two-valued logic, be a strict contradictory counterpart to A. But because the negation is determined in this manner, as given in column 2, we must take this to be an arbitrary definition. It is guided exclusively (as will be shown) by the specific intention lying behind Reichenbach's schema, i.e.

a. Hübner uses the Latin expression *tertium non datur*—literally, there is no third (alternative)—instead of the more common *Satz des ausgeschlossenen Dritten*, (in English, "law of excluded middle"). In what follows, he abbreviates this simply as *TND*. Here this has been translated throughout as "law of excluded middle," followed by *TND* in parentheses.

to design a logical calculus specifically fitted to quantum mechanics. The same holds for column 3. Reichenbach calls the kind of negation in column 2 "complete negation" ($\overline{A}$) and that in column 3 "cyclical or series [*zyklische*] negation" ($\sim A$).

By means of this schema, propositional operators are then defined that correspond to those of disjunction and implication, just as we might find them in a textbook of propositional logic. This can again be shown with a schema:

SCHEMA 2

A	B	Disjunction $A \vee B$	Alternative Implication $A \rightarrow B$
1.T	T	T	T
2.T	U	T	F
3.T	F	T	F
4.U	T	T	T
5.U	U	U	T
6.U	F	U	T
7.F	T	T	T
8.F	U	U	T
9.F	F	F	T

As is apparent, the disjunction in lines 1, 3, 7, and 9 is in agreement with the common definition. The same can be said for the alternative implication in the same lines. In these cases A and B have truth values of true and false only.

If you now add to this last schema a definition of equivalence like the following: "Two statements are equivalent when they are both true, both false, or both undetermined," then we can set up the following equivalences as tautological and thus constantly true for the system:

(3) $A \equiv \sim\sim\sim A$,

(4) $\overline{A} \equiv \sim A \vee \sim\sim A$,

(5) $\overline{A} \rightarrow B \equiv \overline{B} \rightarrow A$.

These follow since, if you posit the value of A as true in the first equivalence (3), then in accordance with schema 1, we find that $\sim\sim\sim A$ must also be true; if you posit A as false, then it follows that $\sim\sim\sim A$ is also false; and finally if you posit A as undetermined, then $\sim\sim\sim A$ will similarly be undetermined. Hence we find that the equivalence is true in every possible case, thus always true. The same thing can be demonstrated in a similar manner for the two remaining equivalences (4) and (5) if we utilize schema 2 as well.

Now consider the statement

(6) $$A \lor \sim A \longrightarrow \sim\sim B.$$

From (6) it follows by (3), (4), and (5) that

(7) $$B \lor \sim B \longrightarrow \sim\sim A.$$

This in turn implies (6), so that (6) and (7) can be said to imply each other mutually:

(8) $$A \lor \sim A \longrightarrow \sim\sim B \rightleftarrows B \lor \sim B \longrightarrow \sim\sim A.$$

174 Now, with the aid of the first two definitional schemata, we can translate statement (6) into words as follows:

If A is true or false, then B is undetermined. Statement (7) reads: If B is true or false, then A is undetermined.

Such a relation between A and B is, however, precisely what is to be understood by the notion of complementarity in quantum mechanics. For example: If a measurement of the position of a particle has occurred, then statement A—that the particle is located at such and such a position—is true or false. But then statement B—that this particle has such and such a momentum—is fundamentally undeterminable, hence undetermined. Therefore you could also abbreviate (6) to read: A is complementary to B; subsequently (8) would read: If A is complementary to B, then B is also complementary to A. The complementarity is thus symmetrical; and this symmetry (e.g. of position and momentum) is an empirical law of quantum mechanics.

But then we might ask: What is the more precise nature of this so-called three-valued logic, this logic in which the law of excluded middle (*TND*) does not appear? What else constitutes this logic?

By way of answer we would find out that it is made up of a series of arbitrary definitions, which we might also look at as axiomatic starting points that possess in and of themselves no kind of immediate or intuitively evident universal validity. In fact, they are purposively constructed in a particular manner so that in the end certain empirical facts of quantum mechanics can be formulated in a corresponding interpretation by means of them. Thus we are indeed dealing here with a propositional calculus expressly suited to quantum mechanics. But do we do justice to the concept of logic by calling such a propositional calculus a propositional *logic?*

175 Logic is characterized by the fact that it can be axiomatically formulated. You introduce axioms and you derive theorems from these axioms by means of rules. The traditional conception of logic contains the view

that these axioms must express universally valid inferences. An example of this in syllogistic logic would be the mood of Barbara, in propositional logic the inference "if A, then A." Now according to a definition stemming from Leibniz, the universal validity of logical axioms means that they are valid in all possible worlds. The same thing is meant today when it is said that logic has tautologies for its subject matter—axioms that tell us nothing about the world, since what is valid for all possible worlds can tell us nothing about this particular world that actually exists. To the above I would also add the definition stemming from Lorenzen. According to him, logic is the doctrine of those rules which are permissible for any chosen calculus whatsoever. As can be readily seen, this definition is also tied to the traditional concept of logic.

Now the mutual complementarity of certain statements in modern physics is a contingent feature of this world, something which belongs to its manner of being and not to every possible world. Hence rules of propositional systems that can be used to express this particular manner of being are not permissible rules for any chosen calculus whatsoever or for tautologies. Consequently, you cannot call such an axiomatically developed propositional system a propositional *logic* if you acknowledge in any sense at all the criterion for the adequacy of a definition. This criterion states that the arbitrary element in the determination of a concept finds its limit in the universal use of a concept. If for whatever reason one does not recognize this criterion of adequacy, then it is even less permissible to speak, as here, of using quantum mechanics as the basis of the construction of a new logic; for then one could only declare that one has *arbitrarily chosen* to call a particular propositional calculus a propositional logic. The kind of philosophical meaning that such a view lays claim to, some kind of new general knowledge about the forms and manner of thought itself, could obviously never be contained in such an arbitrary explanation.

Moreover, even if we ignore all this, the renunciation of the law of excluded middle (*TND*) which would seem to be suggested by Young's interference experiment and is reflected in the three-valued propositional calculus can in no way be viewed as leading to a modification of the traditionally determined concept of logic. For, as we know today, the form of inference represented by the law of excluded middle (*TND*) is not valid for any chosen calculus whatsoever or in all possible worlds, and hence is not, properly speaking, a fundamental law of logic.

7.2 Mittelstaedt's Attempt

A further attempt to conceive the propositional calculus of quantum mechanics as a quantum *logic* stems from Peter Mittelstaedt and appears in his book *Philosophical Problems of Modern Physics*.[3] There he bases his attempt on the so-called dialogical logic of Lorenzen. The fundamental ideas of this logic can be sketched as follows:[4]

First, it is presupposed that we know how to prove simple, not compound, propositions (propositions like "the moon is round" or "the weather is nice," etc.). Now let us suppose that someone asserts—let us call him the proponent (P)—if A, then B, symbolized here as (A $\rightarrow$ B). Someone else—let us call him the opponent (O)—might wish to contest this claim. Obviously, however, this can only occur if the opponent himself proves A and then demands that the proponent, for his part, prove B, since A $\rightarrow$ B obviously consists in the assertion that if A exists, then B also exists. In the case where the proponent is the winner, we can schematize this fabricated dialogue in the following form:

P	O
Assert.: A $\rightarrow$ B	Assert.: A
How do you know A?	Proof of A
Assert.: B	How do you know B?
Proof of B.	

If the opponent wishes to win, he must first prove A in the hope that the proponent cannot prove B. The opponent loses if either he does not prove A or the proponent succeeds in proving B. And the proponent loses if the opponent provides the proof for A, but he himself cannot supply the proof for B.

Now let us suppose that the proponent asserts the following: A $\rightarrow$ (B $\rightarrow$ A). And the opponent contests this. How would the dialogue develop now? We can again use a schema to show this.

P	O
1. A $\rightarrow$ (B $\rightarrow$ A)	1. A
2. How do you know A?	2. Proof of A
3. (B $\rightarrow$ A)	3. B
4. How do you know B?	4. Proof of B
5. A	5. How do you know A?
6. See O, step 2.	

P would already be the winner in step 2 if O could not supply the proof for A. But since O comes up with this proof, P must assert the conclusion of the asserted implication in step 1. O must then prove B or lose. Since he succeeds in this, P must again assert a conclusion, namely, that contained in the implication (B $\rightarrow$ A). And this works for him since all he has to do is point out that O has himself already produced the proof for A in the second step.

Accordingly, not only has the proponent won, but he will always win this dialogue irrespective of the particular content of A and B and completely independent of whether A and B can in fact be proved. The assertion A ⟶ (B ⟶ A) can therefore be called universally valid, because it can be defended in absolutely any chosen dialogue and it will always come off the winner. It is then precisely for this reason that it is an assertion belonging to logic: or to use Lorenzen's terminology, it belongs to the so-called effective propositional logic (*effektiven Aussagenlogik*), for which the guiding idea is that of universal validity. It is precisely for the same reason that the law of excluded middle (*TND*) does not appear in this logic.

Now Mittelstaedt is of the opinion that in light of quantum mechanics the effective propositional logic is in part either false or not applicable. Here we find then that not only has the law of excluded middle (*TND*) been attacked, but there is also a criticism of that form of logic which has relinquished its claim to this axiom (*TND*) and thus apparently re-established itself as a universally valid discipline. Mittelstaedt writes:

> Either one presupposes a knowledge of quantum theory in the sense that, given two propositions, it is known whether they are commensurable or not—in this case the logic remains valid in its full extent, but some of its laws lose their applicability when applied to incommensurable properties. Or one explicitly excludes a knowledge of quantum mechanics and hence relates all measurable properties in similar manner to a quantum-mechanical system, i.e. one introduces fictional objects—in this case some laws of classical logic become false. The laws of logic that remain valid under these circumstances then form in their totality a quantum logic.[5]

179

But how, then, one might immediately ask, can a part of logic become false simply because we exclude a particular kind of empirical knowledge, i.e. knowledge pertaining to quantum mechanics?

But now let us look more closely at Mittelstaedt's own train of argument as he develops it. In this he makes use of the previously elaborated example of a proposition that is always valid because it will always come off the victor in any dialogue, that is, the proposition A ⟶ (B ⟶ A). Let us assume A and B to be complementary statements of quantum physics. In this case, then, O-2 means that A has been proved by a measurement and O-4 means that B has also been proved by a measurement. But in terms of quantum mechanics, when we come to step 6 the proponent can no longer refer to O-2 because the measurement by which B has been proved annuls the one by which A had been proved, since, in fact, we are dealing here with complementary statements. So, as a result, A is no longer available for use in step 6. Hence the proponent can no longer answer the opponent's question, "How do you know A?" (O-5); consequently he has lost according to Mittelstaedt.

If, therefore, the proposition A ⟶ (B ⟶ A) is simply asserted to be universally and absolutely valid out of an ignorance or an ignoring of quantum mechanics as is the case in this effective logic, then what was said above would prove false.

But things would supposedly turn out differently in a case where knowledge of quantum mechanics is not excluded. In this case, Mittelstaedt maintains, the proponent could defend the proposition A ⟶ (B ⟶ A) in the dialogue because in step 4 the opponent would have annulled his own presupposition; that is, his proof of B would have annulled his proof of A. Thus, viewed in this light, the implication in question would be universally demonstrable on grounds that it would not be applicable at all.

But this conception is untenable for the following reason: If you read the proposition A ⟶ (B ⟶ A) in a way that is guided by the exact definitions in logic, then it becomes clear that its universal validity is assured and that this is in no way dependent upon some kind of knowledge pertaining to quantum mechanics. Seen in this light it reads: "*If* A is proved, *then* if B is proved A is also proved." Hence, if A is not proved, the proposition still remains valid because it only states something for *the* case in which A is proved. And if the proof of A is annulled by the proof of B, then once again we arrive at a case in which the premise is not given, that is, the premise which states that A has been proved. So here again the proposition also remains valid. Thus whether or not a logical proposition is applicable in a given case is of no interest in this respect, since this has no effect on its formal truth.

7.3 Stegmüller's Attempt

Recently Stegmüller also asserted that we must have recourse to a nonclassical logic in order to speak about quantum mechanics.[6] Basing his studies on several works by Suppes,[7] Stegmüller begins from the following thesis:

> Quantum physics contains a paradox of probability theory that results from the application of classical probability theory to this realm. According to classical probability theory, a probability must be assigned to every element of the algebra of events. But in the case of quantum physics single events come to the surface which do in fact have a defined probability, whereas their conjunctions do not.[8]

The reasoning behind this thesis can be presented here in an abbreviated outline which will be sufficient for our subsequent critical treatment.

First, the concept of a "classical algebra of events" must be defined. By this we are to understand a nonempty set A composed of subsets of a set Ω such that, for any a, b $\in$ A:

(1) $\bar{a} \in A$ ($\bar{a} \equiv$ complement of a with respect to Ω),
(2) $a \cup b \in A$.

Next an "additive probability space" (*additiver Wahrscheinlichkeitsraum*) pertaining to a classical algebra of events A can be defined by introducing a probability function P which must satisfy the following conditions:

(3)   P(a) > 0, if a is not the empty set Φ,
(4)   P(Ω) = 1,
(5)   if a ∩ b = Φ, then P (a ∪ b) = P(a) + P(b).

Finally a "random function" x is defined (this function is often called a "random variable", but Stegmüller's arguments against this usage are convincing) such that if, for example, we denote the obverse (heads) of a coin with 0 and the reverse (tails) with 1, and flip the coin three times, we can then formulate the random function "number of heads" as $x (0,0,0) = 3, x (0,1,0) = 2$, etc. Thus we are dealing here with a function defined for the elements of a set Ω, where the values of x are real numbers. From x we can then derive a *distribution function* Γ_x by applying the probability function P to particular sets formed by means of the random function. This can be expressed in the following form:

$$F_x(x) = P(\{\xi | \xi \in \Omega \wedge x (\xi) \leqslant x\}).$$

In this way the quantities of quantum physics can be interpreted as random functions such that the expectation value E of a distribution function F is expressed in the following formula:

$$E(x) = \sum_{i=1}^{n} x_i F'_x(x_i), \quad F'(x) \equiv \frac{dF(x)}{dx},$$

and for which the standard deviation S is given as

$$S \equiv \sqrt{\sum_{i=1}^{n} (x_i - E(x))^2 F'(x_i)}.$$

We can then state Stegmüller's asserted paradox in terms of the above outline as follows:

Quantum physics can be interpreted as a theory concerning probability distribution functions of random functions. Thus the physical quantities are represented by random functions. Now, if x and y are random functions associated with probability distribution functions F_x and F_y, then they should yield the combined probability distribution function F_{xy} given in the following prescriptive formula:

$$F_{xy}(x,y) = P (\{\xi|\xi \in \Omega \wedge x(\xi) \leqslant x \wedge y(\xi) \leqslant y\}).$$

The construction of this expression should be possible since the operations to be carried out within the braces are defined in accordance with the rules of classical logic and classical probability theory. But in quantum physics, *in contradiction to this,* there is no corresponding combined probability distribution function for the single probability distribution functions of particular quantities.[9]

183   According to Stegmüller, there is then only one rational way leading out of this paradox, namely, to redefine the concept of an algebra of events. He does this precisely by stating that the conjunction of two events, a and b, does not exist in all cases. And according to Stegmüller this would mean that the algebra of events, whose elements are still to be thought of as states of affairs (*Sachverhalte*) and/or propositions, would no longer have the structure of a Boolean algebra and that accordingly conditions (1) and (2) could no longer be understood in terms of classical propositional logic and thus could no longer be used to define an appropriate algebra of events. Such a modification, he writes, "in fact amounts to nothing less than the *postulation of a nonclassical logic of events.*"[10]

Now one can argue against this conception in fundamentally the same manner as against that of Mittelstaedt. If classical logic requires that the conjunction of two propositions A and B does in fact exist in a universalizable sense, then this obviously presupposes at the same time that A and B have *definite truth values which are independent of one another.* Thus the rule "A,B $\longrightarrow$ A$\wedge$B" means: If the truth of A as well as the truth of B have both been independently established in their own right, then the truth of A $\wedge$ B is also established. And again the validity of this rule remains unaffected even when the above conditions do not apply.

First of all, then, we should point out that in his discussion of quantum logic Stegmüller, along with Suppes, proceeds on the basis of an extreme or radical view of the uncertainty relation, according to which the measurement of momentum absolutely precludes the possibility of determining a "definite truth value" for any statement concerning position, and vice versa. But if this is the case, then, according to Stegmüller's own presuppositions, that paradox from which he deduced the necessity of

184   maintaining a nonclassical logic of events does not even exist. For, if in dealing with two possible probability distributions A and B we can never ascertain more than *one* fully definite truth value determination, then it follows that there is no formal contradiction pertaining to classical logic when there is no combined probability distribution for A and B taken together.

Hence it seems to me that the expression "quantum logic" is misleading and only serves to confuse matters. Quantum mechanics has not led—as we so often hear today—to a new logic; it has not provided us with insights into new forms of thinking; nor has it hurled logic headlong into the turmoil of the constant flow and progression of the empirical sciences. Rather the case stands just the opposite: Quantum mechanics itself presupposes the universally valid propositions of "effective logic."

With regard to the standpoint in question here, it is then extremely important to bear in mind the reasons that led (for example) to Reichenbach's development of a propositional calculus suited to quantum mechanics, his "three-valued logic." In doing so he proceeded on the basis of an interpretation of quantum-mechanical events arrived at by the Copenhagen school, as founded by Bohr and Heisenberg, which implies the following theorem: If two statements are complementary, then at most only one of them can be meaningful, while the other is meaningless.

This theorem is then taken here to be a physical law—that is, it is taken to be merely another formulation of Heisenberg's uncertainty relation, which does indeed exclude the possibility of a simultaneous measurement of noncommutable quantities. But in this the law would then have taken on a semantical form, in fact purporting to express something about the sense (*Sinn*) of statements; and as such it would thus belong to the metalanguage of quantum mechanics. But there remains something unsatisfactory, even unnatural, in this. Laws are normally formulated in an object language (*objektsprachlich*). In addition the theorem mentioned above refers to the entire class of propositions constituted by meaningful as well as meaningless statements. But then, since it is meant to express a law, in a certain sense it would mean that physics must also include meaningless statements.

Thus we can see that Reichenbach created his so-called three-valued logic *for the sole purpose* of formulating the uncertainty relation within the framework of an object language. Let us then again consider the assertion $A \lor \sim A \longrightarrow \sim \sim B$. Metalinguistically interpreted, this does indeed mean: If A is true or false, then B is undetermined. But it is also the expression of an object language and as such can be read: A or cyclical (series) not A implies cyclical (series) not not B.[11]

In this way we come to see the nature of the real intention behind this so-called three-valued logic. It aims at nothing other than a kind of formulation of quantum-mechanical laws in complete accordance with that formulation which is customary in physics.[12]

185

PART 2 THEORY OF THE HISTORY OF SCIENCE AND OF THE HISTORICAL SCIENCES

8

FOUNDATIONS OF A UNIVERSAL

HISTORISTIC THEORY OF THE

EMPIRICAL SCIENCES

Our age is called the scientific-technological age. By this it is meant that today science plays the decisive role and that almost everything falls under its influence. The position of science today is thus thoroughly analogous to that of theology in the past, even though this might appear something of a strange statement at first glance. Just as theology once permeated the whole fabric of life, and in the final analysis everything was granted meaning, made conceivable, and mastered by means of it, so today we hold science to be competent in all matters and grant it the right to speak on behalf of everything. Earlier the priests were consulted before important undertakings, now the scientists. In public as well as private affairs they are brought in as advisers; this is true even for such cases—indeed especially for such cases—where these scientists represent fields of research which are still highly controversial, like sociology and futurology. Fantastic sums of money are spent on the sciences and allotted for scientific projects, sums comparable to those spent on the construction of cathedrals in the past. And just as it was previously thought that no one could attain salvation without theological guidance, so today we believe that no one can be happy (*glücklich*) without a university education.

To what, then, does science owe this position of overwhelming power? It has its roots in that opinion, born in the Enlightenment, that science and science alone can open up the correct pathway to truth and that it has already attained truth in some areas, or at least draws nearer to it day by day. In the first chapter we found an example of this view in Kant. The picture of nature painted by science is thought to become increasingly more exact and increasingly more encompassing. The statements and theories of science are justified by means of either objective facts or necessarily valid fundamental principles; and by means of these it is constantly checking itself.

It is of little or no concern here that some of the philosophers of science sympathize with empiricism and stress facts while others sympathize more directly with rationalism and call more on fundamental principles, since in the end empiricism and rationalism alike contribute to that trenchant scientific optimism which has been one of the major causes of the tremendous changes to which the world has been subjected since the Enlightenment. Nevertheless, as the previous chapters show, this optimism rests on an illusion.

If we mean by these fundamental principles—as indeed on closer inspection we must—on the one hand those precepts belonging to the categories developed in terms of the theory of the natural sciences as presented in chapter 4, and on the other hand those precepts which play a similar role in other empirical sciences (these among other things will be dealt with in chapter 13), then we are compelled to adopt the following general position: There are no absolute scientific facts or absolutely valid fundamental principles upon which scientific statements or theories can be based in a strict sense, or by means of which such statements and theories can be justified with necessity. Factual assertions (*Tatsachenbehauptungen*) and fundamental principles are, entirely to the contrary, merely parts of theories: they are given within the framework of a theory; they are chosen and valid within this framework; and subsequently they are dependent upon it. This holds for all empirical sciences—for the natural sciences as well as those pertaining to history.[1]

But then this leads to an even more problematic state of affairs. If scientific facts as such are unavoidably dependent on theories, then we must further conclude that they change when the theories change. Therefore it is a mistake to think of science as something which necessarily and constantly improves its knowledge of the selfsame object in the course of its development. We must not let ourselves be deceived by the mere fact that the same words or terms often remain in use when one theory is replaced by another on the supposition that the latter is an improvement on the former. For example, we find expressions like mass, momentum, velocity, time, space everywhere in present-day physics just as in older forms of physics; but they often mean completely different things in accordance with the theoretical context within which they are used, whether that be the Cartesian, the Newtonian, or the framework of Einsteinian physics, etc.

For this reason new facts never really crop up on their own; rather it is always the case that they are only discovered in the light of a new theory (which must then have preceded them). It would be futile to try simply to graft new facts onto the previous stage of science, since these facts must be conjoined with the altered context which made them possible in the first place. The older facts are partially reinterpreted in the light of new theories, they are partially bracketed, and they are partially declared to be sheer deceptions. As an example, we can look at the rise of classical mechanics in the seventeenth century. After its fundamental idea had been set forth, this theory allowed for the discovery of completely new laws of motion. Following this, the appearances, which had previously been interpreted in accordance with the more biological view of Aristotle, were now conceived in a completely new way and everything was seen from within a mechanistic framework. The organic framework that had once been dominant had hardly any place at all in this new conception. "Animalia sunt automata," Descartes declared categorically, and with this expressly swept the older view from the table. This example clearly shows, I believe, that what is new in the sciences cannot simply be viewed as an extension, improvement, or enlargement of the old. Often the occurrence of the new is more readily analogous to the genesis of a completely differently interpreted world, the contents of which are changed so as to be partly more extensively developed and partly more narrowly viewed.

On the basis of the results of the previous chapters, I would like to summarize what is involved here one more time. The optimism of the empiristic-rationalistic standpoint of science is an illusion for the following reasons:

1. There are no absolute scientific facts or absolute fundamental principles upon which science can be based.
2. Science does not necessarily present us with a constantly improving and ever-expanding picture of the same objects and the same empirical content.
3. There is not the least reason to suppose that science is drawing nearer in the course of history to some absolute, that is, theory-free, truth. (The idea of absolute truth will be more extensively discussed in chapter 11.)

8.1 A Historical Situation Determines What the Facts and Fundamental Principles Will Be, Not Vice Versa; Historical Systems and Historical System-Ensembles[a]

In the previous chapters we have shown, mainly through examples, that statements pertaining to the natural sciences can admit of a historical justification. The present chapter should serve, *first,* to universalize these results and thus to extend them to the level of all empirical theories whatsoever, whether or not they are theories of the natural sciences; *second,* to analyze the *logical structure of this form of justification* more precisely and systematically, and to elaborate the categories necessary to this task; *third,* and finally, to allay, by showing it to be unjustified, the fear that such a *historistic* interpretation of science leaves the door open to relativism and skepticism, as well as destroying any hope for scientific progress.

a. The German terms used here are *geschichtliche Systeme* and *geschichtliche Systemmengen.* The first is straightforwardly translated as "historical system(s)"; the second has been translated here and throughout as "historical system-ensemble(s)." A *Systemmenge* is literally "a set of systems." However, as this represents a rather special kind of "set," it seems appropriate to differentiate this from a purely mathematical concept. To this end the word "ensemble" has been used.

I will begin with the thesis that it is a historical situation which decides what the scientific facts and fundamental principles will be, and not vice versa.

To illustrate this point, I will first define the concept "historical situation" with the aid of two categories belonging to the realm of the *historical sciences*. These are the categories of *"historical system"*[2] and *"historical system-ensemble."*

The category "historical system" relates to the structure of any and all historical processes, thus not merely to those pertaining to science. On the one hand such processes proceed in conformity with laws of nature, such as laws of biology, psychology, and physics. On the other hand, however, they are also in conformity with man-made rules; and here I only want to call attention to this latter element. There are as many kinds of these rules as there are facets of human life. Consider the rules governing the everyday intercourse among people or in general the manifold ways in which people can be related to one another: rules pertaining to the business world, economics, and the state; also the rules of art, music, religion, and of course language. Since on the one side such rules arise historically, and hence are also subject to historical transformation, while on the other they also serve to provide our lives with a kind of systematic constitution, I call them *historical systems of rules,* abbreviated in what follows simply as "systems." Obviously, most of these systems do not correspond to ideals of exactness and completeness; but in general they are nevertheless exact enough to be applied in the situations for which they were devised. Thus contrary to a widely held opinion, we can assert that our extrascientific life also possesses, to a considerable extent, a certain rationality and logic, insofar as this finds its expression within the framework of such systems.

A historical system can be thought of either as an axiomatic system or as something describable by means of such a system. When dealing with an exact axiomatic system, and hence with an *ideal case,* we find a small number of precisely formulated axioms and a deductive mechanism by means of which other statements or symbols can be produced or derived from these axioms. An example of such a system would be a rigorously formulated theory of physics taken as an object for study in the history of science. An example of a system which *is not* itself an axiomatic system, but can be adequately *described* by means of such a system, would be an existing machine for which there is a mathematical model; such a system could serve as an object for somebody writing the history of technology. When the system in question is not ideal—and, as noted, in most cases it is not—then in addition it cannot be described by means of an ideal system. In such cases, even though axioms and deductions might indeed be quite obviously given or describable, and might, as stated, be assessable within the context of given situations, nevertheless the systems in question here remain more or less vague, and at the very least do not admit of rigorous formalization. To this group belong the previously mentioned systems of practical and cultural life, as well as value systems, systems of law and justice (*Rechtssysteme*), and political systems (*politische Kalküle*). All of these, even when they are not more or less vague axiomatic systems, are nonetheless describable as such.

By the second historical category, the "historical system-ensemble," I mean a structured set of systems, in part contemporary (to the time in question) and in part handed down, which are ordered among themselves in accordance with various relations and within whose compass a given society moves at any give time. Scientific systems, that is, theories and hierarchies of theories, as well as the rules governing scientific work, are thus all a part of this collective ensemble that presents us with the world of rules within which we live and act at any given time.

The relations governing the elements of this ensemble might, for instance, be those of practical motivation—e.g. when a system is morally judged, sanctioned, or rejected, in the light of another system. Here I need only remind the reader of former times in which it was customary to work out the correctives to theoretical-scientific statements in terms of theological-ethical axioms, or of the tendency on the rise today of judging scientific projects in accordance with guidelines of so-called social relevance, etc. Another kind of relation between systems concerns the theoretical critique of one system in terms of another. Here we might mention Leibniz's critique of Newton, which utilized Leibniz's relativistic philosophy of space, or conversely Euler's critique of such relativistic philosophies based on his acceptance of the principle of inertia as self-evident. As another example, we might point to the frequent rejection of ethical axioms that has as its basis the theoretical assertion that all events, thus human actions as well, betray a thoroughgoing determinism. Here we will have to be satisfied with these few examples of the possible relations between systems. Before passing on, however, it must be emphasized that within a given system-ensemble we can also find systems that are incompatible with one another and others that are even incommensurable with each other. 196

With the aid of the historical categories just elaborated, I can now proceed to a more precise definition of the concept "historical situation." By this I understand a historical time period (*Zeitraum*) that is dominated by a particular system-ensemble. And I now assert that every historical time period is so constituted.

If we want to say more explicitly that at this or that time there was this or that system-ensemble, then we must assume a logical conjunction of axiomatic theories, each of which is coordinated with only one system of this system-ensemble as its description. Obviously, this is only a *regulative idea;* but I would like to differentiate the present use of the term from its use in the Kantian system. Here I call this idea regulative only in a *practical* sense. Thus it is only an idea because no one can exhaustively describe a time period in the manner indicated; it is a regulative idea because it calls on us to move out from every particular known connection within the system-ensemble to a broader and more comprehensive set of connections; and it is only an idea in a practical sense precisely because, contrary to Kant's theoretical ideas, it only refers to a *finite* set and is thus only practically nonrealizable.

Now, since the ensemble has a structure in virtue of the relations obtaining between its elements, it might be possible to suppose that perhaps all these elements are deducible from one fundamental element of the ensemble. But this is simply not the case. We are indeed able to progress regulatively from any point in the nexus of the ensemble in this or that direction, and thereby to establish several lines of connections, though, generally speaking, we cannot proceed in all directions. But as previously stated, many elements of the ensemble are nonhomogeneous, incommensurable with each other, or just plain contradictory.

Let us then summarize once more: A system-ensemble is describable in accordance with a practical-regulative idea in terms of a conjunction of axiomatic theories. We can construct relations between the elements of this conjunction in the manner indicated above.

Thus, when I spoke previously of a historical situation as that which determines what the facts and fundamental principles will be and not vice versa, stated more precisely I mean that the system-ensemble characteristic of a particular historical time period exerts this power of determination. By way of clarification let us first consider a few examples. These examples have purposively been drawn from the previous chapters in order to indicate how they fit into this new interpretational schema and how this schema can be used to treat them.

The facts and principles underlying the Ptolemaic system, as we pointed out before, were mediated by a particular interpretation of the dominant Aristotelian doctrine of the day which differentiated between the translunar and sublunar spheres. According to this doctrine, human perception is a dependable source of knowledge only within the earthly realm. Viewed in this light, the facts pertaining to the heavens do not contradict Ptolemaic astronomy; in fact they can actually be seen as being in agreement with it. Moreover, this astronomy is based partially on the fundamental principles of physics, metaphysics, and theology that were in force at the time.[3] In addition, we observed that Einstein believed reality to be composed of substances possessing certain intrinsic properties, properties that remained completely unaffected by the various manners in which these substances were related to one another.[4] This view finds its origin in a conception handed down from ancient philosophy, a view which received its most essential formulation at the hands of Aristotle and Descartes. On the other hand, as we also saw, Bohr believed that reality was essentially composed of relations between substances; this view in turn was chiefly mediated and influenced by the dialectical philosophy of Kierkegaard and James. The disagreement between Einstein and Bohr shows quite clearly that "facts" have a different meaning for the two of them and that for them facts are not given in the same manner. Consequently, Einstein rejects quantum mechanics as incomplete because it leaves out a great deal of what Einstein himself considered to be pertinent to the notion of a fact whereas, at the same time, Bohr denies that "facts," in Einstein's sense, are facts in any sense at all.

Let us also take an example from the historical sciences (these sciences will be treated more thoroughly in chapter 13), namely, the theory of facts disseminated by the school of historical positivism. The leading exponents of this school are the American scholars Andrew D. White, John Fiske, H. B. Adams, Walter P. Webb, et al. In a radical development and extension of a number of rather misunderstood ideas appropriated from the German historians— e.g. von Savigny, Niebuhr, Lachmann, and Ranke— they became convinced that there were absolute facts permeating the whole of history and that the proper task of the historian was to research these facts. But in their opinion this could only be accomplished by basing the investigations exclusively on original documents, excavations, ruins, the science of weaponry (*Waffenkunde*), treaties, letters, notebooks and diaries, chronicles, historical writings, etc. Supposedly, then, only the in-depth study of such facts could indicate what actually took place and how it came about. But this theory of historical facts also has different roots. Here we can point to biblical criticism, the methods of classical philology, Enlightenment philosophy, and last but by no means least the stance of the natural sciences. The influence of this final element is echoed in Webb's noteworthy statement that Ranke transformed the lecture hall into a laboratory where documents were used instead of retorts.[5] This entire conception was then later repudiated, especially by the German historical school. Facts, they maintained, have to be interpreted by the historian within the context of his own conceptual designs; hence they never present us with absolutes in and of themselves.[6] Here we can once again clearly see that what is and is not held to be a historical fact is dependent upon manifold theories, all of which find their basis in a historical situation.

8.2 The Development of the Sciences Is Essentially Evoked by Inconsistencies within System-Ensembles; Seven Laws of Historical Processes

Whether we are dealing with Ptolemy, Einstein, Bohr, Webb, or some other figure, we find that they can all be interpreted in terms of the system-ensemble within which they lived and worked, and which is bound up with a particular time period. This ensemble is the very ground upon which we stand, the air we breathe, and the light in which everything is made visible to us.

Having pointed this out in advance, we now encounter the pressing question of the meaning of scientific progress and how one can escape the problem of relativism.

First of all, in the light of what has already been presented, we can see that the development of the sciences is essentially evoked by inconsistencies within system-ensembles and consists in the internal transformation of such ensembles. I will again attempt to illustrate this with an example, and again I will take this example from the previous chapters; but by employing it anew within the present context, I should lend clarity to what we have now gained by way of a more penetrating analysis.

Let us consider the system-ensemble of the Renaissance. As we have already seen, there belong to this (among other things) a certain emancipatory humanism and certain doctrines of theology, Ptolemaic astronomy, and Aristotelian physics. This humanism, which aims at bringing man closer to God, contradicts Ptolemaic astronomy, for which the Earth is coincidental with the place of a *status corruptionis;* and this astronomy was closely tied to the theology of the time. The contradiction was resolved by Copernicus by means of a change in astronomy, a change which was certainly to the advantage of humanism. However, in doing this a new antithesis arose, the antithesis between this new astronomy and the body of Aristotelian physics which still remained intact. So the attempt was made to eliminate this antithesis as well. But when this had finally been accomplished by Newton (at the latest), not only Aristotle but Copernicus had been left behind.[7] At this time the altered landscape of natural science began to react again with humanism and theology, until finally everything had been changed, astronomy, physics, humanism, and theology; thus—and this cannot be overemphasized—along with this, the fundamental principles and factual assertions belonging to these disciplines had also been changed. The result was therefore an entirely new system-ensemble and a completely altered historical situation.

These examples not only show that the concept of the system-ensemble is particularly appropriate to the task of clearly and distinctly conceptualizing, classifying, and ordering the kinds of occurrences depicted here, but also serve to establish the previous assertion that such events have their origins in inconsistencies within a system-ensemble. From its beginning to its end, the Renaissance system-ensemble, like that of every other period, persists in its flawed or tainted character; thus it is always driven on by the attempt to eliminate its flaws. Moreover, this example shows us something else of equal importance: It indicates that this "catharsis" can only succeed on the basis of available means, that is, on the basis of means which are thus made available by the system-ensemble itself. Solutions are sought within the given situation, a situation which transforms itself purely on the basis of itself; and it is precisely this kind of transformation that I refer to when I speak of an internal transformation of a system-ensemble. This becomes quite clear when we answer the question of what was actually done in the attempt to eliminate inconsistencies: A decision was made in favor of one part of the system-ensemble; this was followed by the attempt to assimilate the other elements under the one chosen element.

Critique and creative transformation are thus both grounded in what is already historically on hand. To all of this we can add one final point of great significance: In the case under consideration, those determinant or constitutive elements of the system-ensemble which triumphed were actually in much greater contradiction to the factual assertions they brought along with them than was the case with the other elements of the system and their factual assertions over which they triumphed. The rotation of the Earth remained an unsolved mystery until the law of inertia was discovered; correspondingly the ad hoc physics invented to go along with the Copernican theory lagged far behind the Aristotelian physics it replaced. Hence it was not the discovery of new facts but rather the internal inconsistencies of the system-ensemble which served here as the mainspring for change. I should like to depict this in the following figurative or metaphorical statement: *The movement of the sciences is essentially a self-movement of system-ensembles.*

We must hasten to add that this has nothing in common with Hegelian philosophy, though at first glance it might appear to be quite similar. But as it would take us too far afield to go into particulars, a few salient points will have to serve here to differentiate the two. The inconsistencies I have in mind and the processes which bring them about are not dialectical in nature. The emancipatory humanism of the Renaissance, for instance, is not related to Ptolemaic astronomy as thesis to antithesis in the sense given this by Hegel, since here there can be no talk of the one being driven forth out of the other with necessity. Indeed, neither the disharmony of systems nor the solution to this can be conceived in terms of strict rational necessity, because the elements that come into contradiction, or which are supposedly brought to resolution, do not for the most part occur in themselves in a strictly unambiguous manner. Even scientific theories when set over and against extrascientific systems of rules rarely prove an exception in this regard and only differ from the latter in their degree of exactness. The reason for this, however, is not to be found in any kind of sloppiness on the part of the theory, but rather is directly related to the necessary rejection of formal perfection, which proves too lethargic and sterile to keep pace with the constantly changing nature of situations. Thus systems, and this means scientific systems as well, do not generally possess any rigorous sense of closure (*Geschlossenheit*); rather they are merely cut to order for the uses to which they are put at a given time. Influenced as systems are by this need to accommodate themselves to change, it is not always possible to determine, in a strict or rigorous sense, which conclusions should be drawn regarding a given system. This means that a certain amount of freedom remains for the construction and interpretation of systems, which in turn makes it impossible to conceptualize the inconsistencies between systems and their resolutions with any strict rational necessity. On the other hand, the Hegelian dialectic, if such were possible, would have to be nothing less than a process of thought thinking itself as thinking, the necessity, rigor and precision of which being that accorded formal logical insights. And for Hegel this demand for necessity and logical rigor in history is made even more requisite, since the system is consecrated and sanctified by the world-spirit. However, I am able to find nothing of this in historical occurrences and processes.

Thus, in opposition to Hegel, I stress the *contingency* in history. For one thing this contingency pertains to the spontaneous acts which serve to bring the previously mentioned obscurities in the practical employment of systems into more or less clear contradictions as well as leading to their solutions. I call these acts "spontaneous" because there is nothing about reason itself that compels us to follow any *one* course of possible action. Furthermore, everything empirical is also contingent. The empirical is not at all eliminated because of the dependence of facts upon theories (cf. chapter 3). To the contrary, every system-ensemble is precisely a concrete manifestation of those very possibilities which enable people to envisage or say anything about reality at all. As such the system-ensemble exhibits, to use Kant's terminology, "the conditions of the possibility" of the production of experiences. These conditions change historically (cf. chapter 4), and herein lies the difference between this theory and Kant's. But the manner in which reality as a whole shows itself under the conditions of a given system-ensemble cannot be predicted with necessity; hence this is also something contingent, just as are the reactions to this view of reality, which are determined by the same system-ensemble.[8]

At this point I would also like to make mention of the opinion that historical processes are determined by nature, that is, by psychological, biological, and physical laws, etc. In this respect some people point to the feelings or emotions that have always affected man, such as love, hate, vengefulness, and vanity, and to drives like hunger, thirst, and sex. All of this is in turn to be viewed in connection with climatic conditions, geography, and the like. Now it follows from the immediately preceding remarks about the role of experience that my metaphorical depiction concerning the self-movement of system-ensembles cannot be understood as involving a denial of the influence of such constant and thus nonhistorical natural factors, or as something which leaves no room for their efficacy. But it does seem to me that even these factors can become effective only within a system-ensemble, and that the necessary conditions and contents which make this possible are only present within such a context.

Accordingly, something like the sensual drive of Salome to give herself to Jochanaan is permeated by Jewish pre-Christian metaphysics. The pederasty prevalent in antiquity also shows that even the sex drive can be channeled into culturally determined paths. Werther's love is inextricably bound up with the sentimentality of the Sturm und Drang period, while that of Tristan is tied to a medieval or, in Wagner's interpretation, to a Schopenhauerian mysticism. The pistol shot of an assassin is indeed a physical occurrence, but no Brutus could have pulled the trigger, just as no Roman could have become mentally fatigued from driving too long on the Autobahn.

In accordance with these preliminary considerations, we can now formulate some general structural laws of history that are exemplified by the events of the Renaissance depicted earlier:

1. Every historical period is determined by a system-ensemble.
2. Every system-ensemble is internally inconsistent and unstable.
3. All system-ensembles change as a result of the attempt to eliminate such inconsistencies.
4. This takes place by means of the accommodation of one part of the system-enemble to another part.
5. This occurrence is not strictly determined.
6. The limits of the determination are set by the free play that allows for the vagueness of the systems.
7. Every historical event occurs within a system-ensemble, even though it is codetermined by natural factors; and nothing can cause completely foreign elements to be introduced into this, nor can elements be entirely dropped or lost. (We should also add that we are dealing here with an idealization to the extent that we have disregarded exchange with foreign historical systems and cultures.) 206

However, it is necessary to supplement these laws with an important comment. For we must heed the fact that these laws are grounded on a *purely logical analysis* of science and the manner in which science regards its own history as well as any history whatsoever. To a certan extent this point of view grows out of the observations made in chapter 3, which serves as the basis for the present section (as well as others). There we expressly pointed out that the treatment had a purely logical analysis as its object; that this also holds for what has been said here will become the matter for a more extensive treatment in chapter 10. But for the present we might only note the following by way of explanation: *The laws introduced here are not representative of a particular empirical theory of history; rather they deal with universal a priori principles which science as such must avail itself of if it is desirous of describing and conceptualizing history by utilizing the theoretical means and categories belonging to it (i.e. systems and system-ensembles).*

The nature of these historical structural laws might be made particularly clear by comparing them to the following structural law of nature, which can now be safely simplified to read: Nature is a system of causal laws. This statement as well might simply be read as referring to an a priori principle pertaining to any scientific treatment of nature whatsoever, and thus not as the axiom of a particular theory concerning a particular system of nature. Thus here, just as in the preceding passage, we are concerned with a priori schemata of the scientific manner of viewing things, that is, with the possibilities of any kind of scientific experience at all—and this applies both to the historical sciences on the one hand and to the natural sciences on the other.

8.3 A Historistic Mode of Treatment
Is Not Necessarily a Relativistic One

Thus the present set of structural laws does indeed say something about the incessant internal transformations of system-ensembles; but this does not answer the questions concerning progress and relativism which are unavoidably posed in this context. I will first turn to the problem of relativism.

The point of view associated with relativism maintains that the only operative force in play in decisions of truth and falsity, or in determinations of what is good or bad, is one of arbitrariness and accident or a kind of fate implicit in history.[b] However, nothing of this sort is necessarily implied if we proceed on the basis of the structural laws mentioned above.

First of all, what is true or false, good or bad, in terms of systems is *founded* within the systems and not in some kind of accidental notion or something woven into the fabric of fate. Hence, within these systems, truth and falsity, etc., can be determined. In addition, there are also rational grounds for systems on the whole and for their transformations within a given historical situation.

If, for instance, we begin with the view that space is Euclidean in nature (an axiomatic precept in terms of the categories given in chapter 4) and if we have further laid down what the nature of an observation is, what a fact is, a confirmation, a falsification, etc. (judicative precepts in terms of chapter 4), then by utilizing these presuppositions, and given particular circumstances, we will make the discovery, and thereby recognize the

b. Hübner uses the Latin *Fatum*. It is necessary to remember in this context that the ancient notion of *Fate* often combines the apparently contradictory concepts of determinism and capriciousness (fickle fate).

truth of the view, that space is full of gravitational forces. These presuppositions for their part, however, were historically given neither arbitrarily nor fatalistically; rather we can now say that they were founded within the system-ensemble and developed out of the rationalism of Renaissance humanism and its underlying principles. It is senseless for us to say today that space either *is* or *is not* Euclidean; but it is highly meaningful to say that the supposition of its Euclidean nature was a well-grounded and decidable part of the given conditions belonging to the Renaissance. These are conditions that we no longer find present, for which reason the question about the nature of the universe is now posed in a completely different way.[9]

Perhaps a comparison might serve to clarify matters yet further. Let us assume that some people are playing cards. The rules of the game will then determine what is true or false, good or bad. It may be true, for instance, that a person will lose when a particular suit is trump and he does not have a single trump card. It is taken to be good tactics to bid overcautiously rather than too rashly, etc. Now assume further that the players come to see that there are certain inconsistencies in the rules. They will change the rules; along with this what is true and false, good and bad, in the game will also be changed. After a certain time has passed, these new rules might also appear to be unsatisfactory and the players will change them again; again this will have the effects mentioned above; and the whole process goes on and on. Now it is easy to imagine that in the end they will be playing a game that has very little in common with the original game (even though they might perhaps still call it by the same name). It would be rather senseless to bring in such an example if by means of it you wished to show what relativism is; for not only, if I might be allowed to use such terms, do we find a kind of "situational logic" here, but moreover the very shifts in situations spring from a certain logic as well. Now for our part we are playing the game of experience, a game where the results are more or less necessitated and in which the conditions are repeatedly being changed, even though the changes are grounded.

Here again I would like to lay stress on the fact that I am not a Hegelian and I am in no way maintaining that the history of the sciences reveals a strictly logical progression, even in the modified sense of the progression of a strict situational logic. In point of fact this history reveals nothing of the kind. Nevertheless, I wanted to point out that a consistent historical standpoint cannot simply be equated with relativism, unless the word is used to describe a state of affairs in which we have lost the fear of subjective caprice and historical fatalism.

At the outset I stated that there is not the slightest reason to postulate an absolute truth we are continually approaching, since we know of no absolute facts and no absolutely valid principles that might serve to point the way to such a truth. Nor is it the case that *the same* objects are

209

210

constantly being dealt with in the advancing course of research in an increasingly more adequate manner. New horizons, as it were, continually come into view and recede again, horizons which offer us entirely new and different vistas and experiences. These horizons possess a knowable and applicable relation to a given situation, but bear no relation to some imaginary absolute truth. (Again, more will be said about this in chapter 11.)

Therefore I believe we should give up once and for all the idea that we can compare the process of development belonging to scientific knowledge with the painting of a portrait, the details of which can be progressively filled in, and this in an increasingly adequate manner, so as to resemble the actual person more and more precisely.

But along with this rejection, we are then presented with the problem of understanding what progress in the sciences can mean in the light of these structural laws.

8.4 Explication and Mutation of Systems: Progress I and Progress II
Two fundamental forms of development can be clearly distinguished here: first, the *explication* (*Explikation*) of scientific systems; second, their *mutation* (*Mutation*).[10] By "explication of systems" I understand the formation and evolution of systems insofar as nothing in their foundations is altered, thus, for example, what Kuhn calls "normal science"[11]—i.e. the derivation of theorems from given axioms, the more precise determination of the requisite constants operative within the framework of a theory, etc. A mutation, on the other hand, occurs when the foundations of a system are themselves altered (e.g. the transition from one kind of spatial geometry to another). Accordingly, progress is possible only in terms of these two fundamental forms of historical movement; therefore two fundamental forms of progress must also be distinguished, which I would now like to call Progress I and Progress II, respectively (*Fortschritt I und Fortschritt II*).

But then, when are we able to speak of Progress I in conjunction with an explication or Progress II in conjunction with a mutation? I hold that an explication marks a kind of progress in science inasmuch as it brings to light everything that lies hidden in a system, what it can and cannot accomplish. Indeed, one can say that explication is the foundation of all scientific progress whatsoever, since without this everything remains fragmentary, in mere outline form, and only half complete. Consider the explication of the relativity theory as an example. This explication finds its historical point of origin in the formulation of laws which are covariant for all inertial systems; and this results in the definition of particular concepts, which in turn leads finally to the recognized energy-mass equation. It is astonishing to observe the cosmos which slowly came to light in terms of this theory and to watch how the theory continually kept conquering realm upon realm. In every new prognosis, whether it concerned the perihelial movement of Mercury or the deviation of light rays in the gravitational field of the sun, we find ourselves confronted with notions which ultimately stem from an explication of the original statement of the theory.

Obviously, however, an explication in and of itself is not yet sufficient to speak of the kind of progress involved here, Progress I. For this we 212
must compare the explicated system to other systems, and this we must do with a view to determining its function and meaning within the context of the existing system-ensemble. Only in this way does it become possible to judge whether the rewards have generally been worth the effort, whether the system is to be judged as somehow unfruitful, backward, provincial, obsolete, etc. With regard to this we might also consider the extreme case of the insanity which engenders self-enclosed systems, systems whose trademark is precisely to be found in their idiosyncratic stance within the existent intellectual framework. But then in what can the function and meaning of a scientific system within a system-ensemble consist, such that it would permit its explication to become the kind of development representative of Progress I? In order to answer this question, we must first consider the nature of Progress II, something which must have at its basis a mutation.

8.5 Progress I and Progress II Are Based on a Harmonization (Harmonisierung) of System-Ensembles

A mutation in and of itself can no more be regarded as a synonym for progress than an explication, and this for reasons analogous to those given in the previous section. Nobody would regard a mutation as progress if it were based (e.g.) solely on such things as sheer arbitrariness, addiction to innovation, pomposity, or utter insanity, just to mention a few. But if this be admitted, then there is indeed nowhere else to turn in the search for rational arguments for a mutation except to the given system-ensemble. Again, there is no nonhistorical source existing outside this framework to which we might turn in order to find criteria by which to measure progress. But then if this is so, if we are bound to remain within the framework of a given system-ensemble, if there is no way leading beyond it, and if then all change is only to be produced out of the system-ensemble 213
itself, then the reason for such change must obviously reside fundamentally in the fact that the internal agreement of the system-ensemble will actually be promoted by such change. And this is to say that we can judge a mutation to be progressive only to the extent that it contributes (1) to the elimination of contradictions; (2) to the elimination of elements of unclarity; and (3) to the creation of the most comprehensive, most internally consistent system of interrelations possible. Such contributions I

entitle *harmonizations of the system-ensemble* (*Harmonisierungen der Systemmenge*).

Here again we can take the relativity theory as a paradigm. At the basis of Einstein's decision to hazard a mutation with his special theory of relativity lay the intention of reconciling Maxwell's theory of light with one of the major principles of classical physics—the equivalence of all inertial systems. When he later determined that this reconciliation could only be had at the expense of the exclusion of the law of gravity, he carried out the second mutation of the system, that mutation which led to the general theory of relativity. He himself expressly confessed that the idea of the harmony of the universe served as his guiding light in the formulation of this theory. I would put it a little less speculatively: in truth he was guided by the idea of the harmony of the scientific system within the framework of the system-ensemble present to him.

With this in mind, we can now answer the previously posed question concerning the kind of function and meaning a system should possess in order for its explication to be viewed as Progress I: It must contribute to the harmonization of the system-ensemble just as the mutation which has produced it.

Once again let us recall the example of explication in the case of the relativity theory. There the explication led to the resultant possibility of bringing a manifold of appearances and principles into a harmonious interrelation which allowed for unified explanations; and this obviously presents us with a contribution of the requisite type. (This point will be discussed more thoroughly in chapter 10.) Conversely, then, an explication in the form of mere critique, one which uncovers contradictions, can also be called progressive in that it obviously carries with it the demand to eliminate these contradictions.

As has become apparent to me through many discussions, the use of the concept "harmonization of a system-ensemble" lends itself to misunderstanding. It is often erroneously understood in an aesthetic light, even though it is intended to be a logical concept, as the criteria enumerated here indicate. It is also occasionally imputed to be a means of justifying the construction of a uniformity that is helped along by omissions and suppressed, or simply counterfeit, elements pertaining to the more inconvenient parts of a system. For example, the following question has been posed to me: "Is not Lysenko's notorious biology also a harmonization of a system-ensemble, namely, the system-ensemble of Soviet socialism, inasmuch as it is fitted to the materialistic principles of this system, even though it disregards scientific methods and experimental results?" The response to this is that in such cases not a single contradiction is actually eliminated; at most they are covered up, if not completely suppressed by force or sheer fraud. The existing biology is so far superior to the so-called Soviet dialectical materialistic view both in clarity and comprehensive internal consistency that even despite all its flaws there can be no doubt, with respect to a confrontation with Lysenko's system, which of the two is to be given preference. Hence, within the present context, "harmonization" as progress means a *genuine* overcoming in thought of the existing difficulties pertaining to a subject, and not some kind of merely apparent or coerced solution. 215

But now let us leave this and turn back once again to the case of Copernicus in order to gain further insight into what is involved in this notion. In the case of Copernicus, as we indicated earlier, we are dealing with the elimination of a contradiction, namely, the contradiction between the humanism of his time and the existing astronomy. He sought to resolve this contradiction by altering astronomy in a way favorable to humanism. Why did he not make the attempt in the opposite direction? Is it not true that this harmony was achieved rather violently insofar as in some other respects the degree of disharmony was actually increased? Copernicus and his successors were after all locked in a dire struggle with the immanent facts of the system! But the Copernican decision in favor of humanism only reveals its rationality and its contribution to the harmonization of the situation if you consider it from the somewhat broader perspective of the system-ensemble "Renaissance" than that represented by astronomy and physics alone. In this broader perspective we see that Renaissance humanism was only a part of a more comprehensive and comparatively more consistent context under whose banner the entire world had begun to transform itself. The discovery of new continents and new seas had led to tremendous changes in trade and travel, changes which eventually shook the very foundations of the hitherto entrenched "sacred" structures of society, like those belonging to the Holy Roman Empire. The secularization of the state began; printing presses and the rise of the middle class destroyed the old hierarchies and privileged classes, thus creating a new individualism. And out of this train of events the idea arose that the Divine Creation, like the construction of a great cosmic machine, had to be understandable by and through human reason.

Thus on the one hand the world was ruled by a multiplicity of systems 216 that hung together for the most part and represented an order which was more or less internally consistent, while on the other hand there were various other systems that not only contradicted this but were struggling as well with an ever increasing number of internal contradictions. Within this framework the Copernican revolution becomes intelligible; and we also come to see that the contradictions that accompanied it were not weighty enough to tip the scales in the other direction. We even find— and this must be emphasized—a progressive element in the fact that the opponents of this Copernican view never tired of pointing out the contradictions inherent in it, since it cannot be denied that Copernicus himself tried somewhat too hard simply to conceal them. Hence it is unjust and false to dismiss the Church, which fought against him, as somehow only backward by comparison.

Thus this goal of the harmonization of a system-ensemble, when posited at all, is not to be limited merely to the sphere of scientific progress; rather, as we have seen, this system-ensemble encompasses much more than science alone. Further, we can now say in general that, contrary to the normal conception, all progress, wherever it may occur, cannot be considered in reference to some extrahistorical end or some kind of *eschaton*—as there is not the slightest foundation for this—nor can we look for such an end to be defined in terms of the total transformation of the system-ensemble, that is, in terms of the creation of something entirely new—for a transformation that does not aim in some way or other at the harmonization of the existing state of affairs can only end in a mentally deranged state of idiosyncrasy.

Thus, while progress might well indeed imply such things as inconsistency, struggle, contradiction, absurdity, or challenge, it only truly becomes worthy of its name when these elements are confined to a somewhat 217 narrower context within which they are unavoidable, whereas in another far broader and more significant context they are outweighed by a general gain in terms of the internal consistency of the system-ensemble as a whole.

From all this it then follows finally that progress, in the sense in which it is understood here and again contrary to the normal conception, is not limited merely to a supposedly "progressive age." Such a belief would mark the height of narrow-minded historical blindness. It is rather the case that every historical system-ensemble permits a harmonization, just as every one of them can be brought to wreck and ruin by a hopeless intensification of its existing inconsistencies.[12] The course of history offers us an abundance of both. Progress I and Progress II are thus normative criteria by which we can in any case assess the worth of explications and mutations, and this not only for the sciences, but for any and all historical systems.

8.6 Neither Progress I nor Progress II Exhibits Constant Development
In the face of all that has been said could anyone still think of progress as some kind of continual, ever-advancing process? Could anyone think that in the course of history every system-ensemble gives way to one which is more harmonic?

Whoever would answer these questions in the affirmative would have to ignore the fact that with the elimination of inconsistencies the same 218 system-ensemble does not always become more and more harmonic and stable. In fact, as I have tried to show by example, in such cases the system-ensemble is altered in its entirety and indeed gradually becomes something completely different, something which thrusts new questions into prominence and offers answers that would have remained inconceivable within the framework of the previous system. Along with this, specific new inconsistencies and difficulties immediately emerge; and we have a completely altered framework.

One could then say along with Wittgenstein that most of the objects that science has dealt with in the course of its history, objects which appear ostensibly to be the same, really bear only a family resemblance to one another. Whether it be space, time, the starry heavens, the forces which move bodies, or some other object of science, we would look in vain for some shared or common meaning which might apply to any of these objects throughout their respective histories and which as such, like a red line traversing the changes in meanings and slowly broadening itself, might serve as the common and continual ground for all the scientific theories devoted to any such object. It was hard enough for mankind to grasp that the same time does not tick off in all parts of the world. It may be even more difficult to grasp that when we investigate some scientific object, both today and as it existed in the past, we are not necessarily speaking about one and the *same* thing. But this must nevertheless be admitted, since there is no thoroughgoing identity that can be maintained here in any strict sense. If there were such, then the so-called essentialist doctrine (essentialism) would have been correct in its assertion that there are essential definitions of entities based on such identities. But try then, if you will, to define concepts like space, time, body, motive force, etc., without taking recourse to the conceptual framework of historically determined theories, without using something which at one time or another was not in fact associated with these concepts, or, in a case where recourse is taken to something which has always been associated with them, without saying something that is merely trivial.

Hence it becomes quite difficult, when dealing with two consecutive   219
system-ensembles, to say that in the final analysis—and I emphasize, *in the final analysis*—the latter is the all-around better because it is taken to be more balanced or simply because it is thought to contain more truth than the former. And this holds no matter how much rational foundation and no matter how much progress is thought to have come into play in each of the questionable mutations leading from the first ensemble to the second, all of which need never have taken place. Progress II, then, is always nothing more than a fleeting piece of good luck (*ein kurzes Glück*), like every other piece of good luck—to say nothing of Progress I, which in the long run always leads of itself to a cessation of progress, and what is more is always brought to a halt by a mutation. Thus progress consists in finding temporary relief from problems only to exchange this respite almost immediately for a new and different set of demands.

What I have sought to do here might thus be viewed as a contribution to the demystification of the sciences as they are understood in the vein of rationalism and/or empiricism, that is, a demystification of the belief in absolute scientific facts and principles. Thereby at the same time I contest the claim that the sciences alone have a virtual monopoly of pathways leading to truth and reality. From the *scientific* point of view the very occurrence and rise of the sciences, together with the correlative truths and realities of these sciences, must be considered as something determined by a historical situation.^c From this it follows that the advance of the sciences cannot be conceived as representative of some kind of *quasi* self-actualization of knowledge, nor simply as the self-actualization of rationality itself. In fact, this process presents us with something that is not essentially different from the rise of the Renaissance ideals, for there is indeed a close connection between the two events. In our scientific-technological world, or, more precisely, in the a priori presuppositions of this world, we have opted for one particular possibility, grounded in one particular situation. We have no cause or inducement for believing that we will always hold to this chosen framework and that we will progress ad infinitum by means of it; nor have we reason to suppose that we would all sink back into barbarity if we were to give it up. To the contrary, as chapter 14 will show, there are various reasons which might lead us to suppose that the ever-increasing number of convulsions endemic to scientific-technological modes of activity, along with the various ideas of progress which belong to such activity, may very well, in themselves, possess a kind of barbarity. But before we consider this, I would like to further elaborate and illustrate the results gained here by turning again to two pertinent examples. The following chapters will deal with these.

220

c. A misunderstanding could arise here which turns of the use of the word *determined* (*bestimmt*). "Determined" should *not* be read in the sense of some kind of *determinism*, that is, as referring to an absolutely "determined" (*determiniert*) order of causal or numerical sequence (as in a physical or mathematical series). The correct reading is much less absolute, as the text makes abundantly clear.

9

THE TRANSITION FROM DESCARTES TO HUYGENS AS SEEN IN THE LIGHT OF THE HISTORISTIC THEORY OF SCIENCE

Since Huygens we have been accustomed to saying that six of Descartes's 221 seven rules of impact are false. On the surface this would seem to be a rather simple finding which can now be written off as a closed case in the annals of history. But in fact, and contrary to this almost unanimously held opinion, we are not dealing here with the simple and straightforward correction and replacement of an error by the truth, but rather, with an occurrence that can be taken to be indicative of precisely that complexity and structure attributed to scientific-historical processes in the previous chapter.

9.1 Descartes's Second and Fourth Rules of Impact as an Example

In order to investigate this matter, let us consider two of the Cartesian rules of impact, the second and the fourth. The second rule states: If two bodies, A and B, approach each other at the same velocity and A is a bit larger[a] than B, then, following their impact, only B will recoil and both bodies will move in the direction of A with the same velocity.[1]

The fourth rule states: If A is at rest and is larger, however slightly, than B, then no matter with what velocity B strikes A, A will never be 222 set in motion and B will be reflected back in the direction from which it came.[2]

Now, even though the second rule might possess a certain plausibility for the mind untrained in physics, everyone will reject the fourth since it contradicts even the simplest experiences. Descartes himself is not bothered at all by this, though he must have recognized it. In connection with the seventh rule of impact, he says something which seems to be valid for him in general: "This does not stand in need of examination, since it is obvious of itself."[3] Thus here he boldly plays off reason against experience; and this is done so provocatively that one should wonder how

a. Descartes generally uses the Latin *magnitudo* for what has been translated here as "largeness." This translation is meant to emphasize Hübner's point that we should not assume a Newtonian concept of inertial mass to be present in Descartes's physics. Hence even "magnitude" is likely to be misunderstood today as a synonym for "quantity," which in turn carries with it a notion of a body with inertial mass. To see how different Descartes's notion is, one must only look at the passages which Hübner quotes on page 128f.

something like this could be at all possible. Indeed, it is surprising that no one ever raised such an obvious question.

It is rather obvious that a modern physicist would not be content to base his rejection of Descartes's views on a study of billiard balls, marbles, and the like, and to let it go at that. Already in Huygens we find someone who brought all his mental powers to bear on the problem and summoned up a considerable theoretical apparatus as a means of establishing the falsity of Descartes's rules of impact. But we see that, in doing so, he also agreed with Descartes's general assessment that mere experience is not quite as self-evident as it might appear at first glance. Thus even though the supposition that the rules of impact are *"per se"* manifest might well be questioned, this in no way gives us the right to assume that they are already *"per probationem"* false.

223   How would a present-day physicist go about checking the two rules mentioned here?

Let us begin with the second rule of impact. First, its premises would have to be translated into the language of mathematics. For "A is larger than B," we write "$m_1 > m_2$," where m_i signifies the inertial mass of the respective bodies. Further, "$u_2 = -u_1$," indicates that the velocities of both bodies *before* the collision are equal but directed contrary to one another. "v_i" stands for the respective velocities *after* the collision. In this way, then, the following two axioms can be formulated:

(1) $$m_1 u_1 + m_2 u_2 = m_1 v_1 + m_2 v_2 ,$$
(2) $$u_1 + v_1 = u_2 + v_2 .$$

From this, as well as from "$u_2 = -u_1$," we then arrive in a purely mathematical manner at the following:

(3) $$v_1 = \frac{(m_1 - 3m_2)}{(m_1 + m_2)} u_1 ,$$

(4) $$v_2 = \frac{(3m_1 - m_2)}{(m_1 + m_2)} u_1 .$$

If we consider the numerators of the fractions in equations (3) and (4), we then have three possibilities:

(a) $m_1 > 3m_2$, (b) $m_1 = 3m_2$, (c) $m_1 < 3m_2$.

If we assume (a) and take u_1 to be positive, then according to (3), v_1 is also positive, thus m_1 continues moving in its original direction after the collision; but according to (4), v_2 will also be positive and therefore m_2 will be reflected back in the direction of m_1. Both of these results are in agreement with Descartes's assertion. But contrary to this assertion, we also find that $v_2 > v_1$, given condition (a), since if we assume that $m_1 = 3m_2 + \delta$ is valid, then by substitution in (3) we get:

$$v_1 = \frac{3m_2 + \delta - 3m_2}{3m_2 + \delta + m_2} u_1 = \frac{\delta}{4m_2 + \delta} u_1$$

and by substitution in (4):

$$v_2 = \frac{9m_2 + 3\delta - m_2}{3m_2 + \delta + m_2} u_1 = \frac{8m_2 + 3\delta}{4m_2 + \delta} u_1 .$$

And this result ($v_2 > v_1$) contradicts Descartes's second rule of impact, according to which both bodies will continue to move with the same velocity after the collision. Correspondingly, it can be shown that in cases (b) and (c) as well we will end up in contradiction to this rule.

Now if we treat the premises of the fourth rule of impact in terms of the same axioms given in (1) and (2) above, we once again end up in contradiction to Descartes's assertion; that is, we find that the larger body at rest will be set in motion in the direction of the smaller moving body after the collision.

The physicist who criticizes Descartes in such a way does not then rest his case, as pointed out earlier, on the simple evidence of everyday experience; rather he confronts Descartes with what he supposes to be *correct* axioms—axioms (1) and (2). Everything else is merely a logical consequence of these axioms and the initial conditions given by Descartes himself (the premises of his laws of impact). Thus Descartes is subjected to the same kind of reproval with which a student might be admonished when, having been given an examination question dealing with the laws of impact in terms of classical physics, he fails to arrive at the answer which is to be expected within this framework. In other words, the physicist believes that he has beaten Descartes at his own game, asserting, as it were, that he should have known better. Such a stance is especially evident in the case of Huygens, since it appears that he was in fact of the opinion that he had refuted Descartes by using Descartes; or otherwise expressed, he was convinced that he had *correctly explicated* Descartes's system while Descartes himself had done this falsely.

9.2 The Meaning of the Cartesian
Rules of Impact: The Divine Mechanics

In order to deal with this critique of the Cartesian rules of impact, we must first recognize that the premises belonging to these rules have been defined here in a completely different manner from that in which Descartes defines them. This critique rests on the tacit acceptance of the presupposition that these premises imply certain statements about the momentum of two bodies (thus about the resultant product of mass and velocity relative to each body) and correspondingly that the law of the conservation of momentum (axiom 1) can be appropriately applied here. But Descartes never even mentions momentum, writing instead about something quite different. He introduces his treatment of the laws of impact in the *Principia,* Pars Secunda, XLIII, 19, with the following words: "Here we must diligently consider the nature of the force by which a body affects another body or by which it resists the effects of another body."[4] And a few lines further on he explains this by stating: "that force must be measured partly in terms of the magnitude of the body in which it is found and the type of surface which separates this body from another, and partly in terms of the velocity of movement and the nature of those contrary modes of movement which bring various bodies into contact with one another."[5] Thus the concept of inertial mass does not appear here at all. But what of Descartes's understanding of velocity? Can we not at least assume that this is the same as that in classical physics?

226   Following his statement that duration is an attribute *in things themselves* (*in rebus ipsis*), Descartes writes: "Of these attributes and modes, some are in the things themselves while others only exist in our thought. Thus, if we distinguish time from duration in general, saying that time is the enumeration of motion, then this is only a mode of thinking."[6] And this he founds on the following: "But in order to measure the duration of all things, we compare their duration to that of the greatest and most regular motions, those which give rise to years and days; and this duration we call time. This adds to duration, taken in its generality, nothing else, then, but a mode of thinking."[7]

Hence duration as something "in the things" is quite different from measured time as something "only in thinking." But upon which of these is the concept of velocity based? On duration or time? Does Descartes use velocity in such a way that it means something inherent in things themselves, thus a *modus in rebus extensis;* or as something pertaining to thinking, hence a *modus cogitandi?*

227   However, velocity is related not only to time or duration, but also to motion. Let us then look at what he says about motion: "But if we would consider what motion means in truth, turning away from its meaning in common usage, then to attribute a determinate nature to it, we can say that *it is the transport of one part of matter or of one body from the vicinity of those bodies that are in direct contact with it, and which are viewed to be at rest, to the vicinity of other bodies.*"[8] Further: "I have finally added that such a transport does not occur from the vicinity of just any contiguous bodies, but rather only from the vicinity of *those bodies which are viewed to be at rest.* Thus the transport itself is recip-

rocal; and it cannot be thought that the body AB is removed from the vicinity of the body CD without it simultaneously being thought that the body CD is removed from the vicinity of the body AB; and the same force and action is required on the one side as the other."[9]

Thus for Descartes motion is something relative. We relate it to something which is *viewed* to be at rest. However, it is always the case that that which is moved can be *thought* to be at rest, just as that which is at rest can be *thought* to be moved. Is it not to be inferred from this that, considered thus for Descartes, motion is only a *modus cogitandi?* And as such must we not also differentiate it from something determinate *in rebus,* that is, from something which is not determined by our more or less arbitrary time measurements or arbitrarily chosen reference systems? 228

One must, I believe, answer these questions affirmatively if Section XXXVI of the Second Part of the *Principia,* that section immediately following the doctrine of motion, is not to remain utterly unintelligible, as it has in all the interpretations that have appeared to date. In this section we find a central element of Descartes's metaphysics. Since he has equated materiality with extension, it can only be set in motion by God; and since God is immutable in that he is a perfect entity, he will maintain the total amount of motion within the universe as constant. The laws of impact are meant to show the details of how this takes place. However, this constancy of motion guaranteed by God would become meaningless if motion could only be determined relatively. In such a case this constancy could *not* be achieved or realized. In modern terms, Descartes's laws of impact, as laws of conservation, could have no validity if they were viewed, for instance, from the standpoint of a rotating reference system. However, if God is the author of motion, then in any case it cannot be relative for him—for God, motion must be *in rebus;* consequently it is a *modus cogitandi* only *for us.*

In this context the following citation from the *Principia* seems to me quite telling: "We recognize perfection in God not only because he is immutable in himself, but also because he only affects things in the most constant and unchangeable manner: This is so much the case that we can attribute no other changes to his creation *than those which present them-* *selves either in clear experience or in divine revelation,* concerning which we either clearly perceive or believe that these changes take place without a change in the Creator; and therefore we cannot infer on the basis of these changes any instability in him. From this it follows with the highest degree of rationality that we must affirm that because God imparted motion in different modes to the parts of matter in their creation, the entirety of matter is preserved by him in the same mode and with the same relations as at its beginning, and thus he preserves the same amount of motion in it."[10] 229

Descartes distinguishes here between changes that are shown by clear experience (*evidens experientia*) and those that are given through divine revelation (*divina revelatio*). Thus, once we come to understand Descartes's disparaging stance respecting experience, a stance which, as previously pointed out, is expressly reaffirmed in connection with the laws of impact, there can be no doubt in determining which changes have been caused by God: the true changes are, namely, those given in divine revelation (*in rebus*), and not the merely apparent ones given through the senses or determined in accordance with some arbitrary relativity as a *modus cogitandi*.

Let us then summarize: The force that Descartes holds to be operative in the interaction of impinging bodies has nothing to do with momentum as we understand it. It relates neither to inertial masses nor to a velocity dependent upon a human time measurement and possible perceptions relating to a body that is only moved in some relative sense. Rather here we see that *Descartes's laws of impact describe fundamental occurrences of nature as if seen from the standpoint of God,* that is, occurrences related to a duration and motion *in rebus* or *sub specie aeternitatis.* Thus these laws are part of a kind of "Divine Mechanics." Consequently, the contradiction which, according to Koyré and Mouy, exists between Descartes's theory of the relativity of motion on the one hand and his divine rules of conservation on the other ceases to exist. We can also understand now that Descartes did not commit an error that could be uncovered and corrected on the basis of simple experience and which he certainly could not have overlooked. Furthermore, we no longer have to consider his reference to the relativity of motion as "cunning tactics," as Koyré has assumed, tactics which Descartes allegedly intended to use as a means of conciliating the Church and reconciling it with Copernican astronomy and the motion of the Earth. Such tactics would indeed have made Cartesian mechanics appear contradictory and obscure.[11] All of these contradictions, difficulties, obscurities, and farfetched hypotheses disappear, however, as soon as we come to recognize that Descartes's laws of impact, as shown, are not based in any sense at all on relative motion as a *modus cogitandi* necessarily determining ordinary experience—a point which seems to have been constantly overlooked, perhaps precisely because it lies so close to the matter.

On similar grounds Descartes also differentiates the Third Part of his *Principia,* "de mundo adspectabili" ("concerning the visible world"), from the Second Part, "de Principiis rerum materialum" ("concerning the principles of material things"), and begins this Third Part with the following words: "Having now discovered certain principles of material things, derived not from the prejudices of the senses but in accordance with the light of reason so that we can entertain no doubt about their truth, we must next examine whether we can explain all natural appearances on the basis of these alone."[12] The invisible world, underlying the visible and alone serving as a ground for the interpretation of the latter, is known by means of an indubitable reason that sees through the sensible to its true cause and knows itself to be one with the light of divine revelation. And it is thus precisely for this reason that Descartes evinces his provocative disinterest in what is clearly perceived by the senses and indeed even challenges this, as is particularly evident in the fourth rule of impact.

9.3 The Internal Contradiction within the Cartesian System

From the preceding it becomes obvious that the critique of Descartes cannot be based on a demonstration that he formulated *false* laws of impact, insofar as these laws are taken to be such as can be pulled into the service of furthering the ends of human experience and seen to rest on the concept of momentum understood as m × v. Again, Descartes never even mentions such things. For this reason a critique can only be directed against *the very fact that* Descartes does not speak about such things, that he concerns himself, so to speak, with "heavenly" rather than "earthly" things. In this context we can then rightly object that the legitimation is missing for the excessive and exalted rationalism to which he thereby abandons himself. That the laws of impact are *clare et distincte* for reason cannot be firmly established, especially since, as Huygens had already noted, they contradict one another in part. But above all we might raise the following objection: On the one hand there is a kind of rationalism in the Cartesian system which, as such, presents itself as seeking practical-technological efficacy and which in fact lays claim to being that system which has established the necessary presuppositions for such efficacy; on the other hand, however, there is a rationalism here which, as an exalted apotheosis of reason, seeks to transcend all earthly matters and present pure theory and pure knowledge in the form of divine revelation. Between these two, then, an unbridgeable chasm can be seen to open up within the Cartesian system. And it is here that we find an inconsistency within the system which is both unsatisfactory and a source of confusion.

As we have already indicated, that form of rationalism which is guaranteed by God, and thus subservient to divine truth, is particularly evident in the laws of impact. On the other side, that notion of rationalism dedicated to practical use finds its most essential expression in the following passage from Part Six of the *Discours de la méthode*. There Descartes writes:

> But as soon as I had acquired some general notions concerning physics . . . I believed that I could not keep them secret, without sinning gravely against the law which obliges us to procure the general good of all men. For they made me see that it is possible to arrive at knowledge which is very useful

in this life, and that instead of that speculative philosophy taught in the Schools, we can discover a practical one, through which, knowing the force and action of fire, water, air, the stars, the heavens, and all the other bodies which surround us, as distinctly as we know the different skills of our artisans, we can use them in the same way for all the purposes to which they are suited, and so make ourselves the masters and possessors, as it were, of nature. This is to be desired not only for the invention of an infinity of artifacts which would allow us the effortless enjoyment of the fruits of the earth and all the commodities that are found there, but especially also for the conservation of health, which is without doubt the primary good, and the basis of all other goods of this life.[13]

234

And in the *Principia,* Pars Quarta, CCIIIf., we find:

Accordingly, just as those who are familiar with automata, when they are informed of the use of a machine, and see some of its parts, easily infer from these the way in which the others, that are not seen by them, are made; so from considering the sensible effects and parts of natural bodies, I have essayed to determine the character of their causes and insensible parts. . . . But here someone will perhaps reply, that although I have supposed causes which could produce all natural objects, we ought not on this account to conclude that they were really produced by these causes; for, just as the same artisan can make two clocks, which, though they both equally well indicate the time, and are not different in outward appearance, have nevertheless nothing resembling in the composition of their wheels; so doubtless the Supreme Maker of things has an infinity of diverse means at his disposal, by each of which he could have made all the things of this world to appear as we see them, without it being possible for the human mind to know which of all these means he chose to employ. I most freely concede this; and I believe that I have done all that was required, if the causes I have assigned are such that their effects accurately correspond to all the phenomena of nature, without determining whether it is by these or by others that they are actually produced. And it will be sufficient for the use of life to know the causes thus imagined, for medicine, mechanics, and in general all the arts to which the knowledge of physics is of service, have for their end only those effects that are sensible, and that are accordingly to be reckoned among the phenomena of nature.[14]

235

Now it seems to me that the latter passage from the *Principia* should not be taken to mean that all those things Descartes presented as absolutely evident principles of nature (the laws of impact for instance) are again to be called into question here as well; rather it seems much more likely that here Descartes is concerned with the more specific discussions presented in the sections dealing with the visible universe and the Earth. But be that as it may, both of the passages cited show quite clearly and precisely the emphatic claim to a science which above all is supposed to serve as a means to the practical-technological mastery of nature. Hence, I repeat again, an unbridgeable chasm opens up between this claim and

Descartes's exposition of the efficacy of a divine mechanics which, as we find it presented in the section on the principles of material things, remains aloof from experience.

9.4 The Transition from Descartes to Huygens
as an Example of the Self-Movement of System-Ensembles

I stated initially that the Cartesian laws of impact and their subsequent critique offer us an interesting example of the complexity and structure of the scientific-historical processes described in the previous chapter. I now wish to elaborate on this point.

First, we can say that the view which originated with Huygens and is still generally held today, namely, that Descartes has been empirically refuted, is both in the service of, and motivated by, the timeworn cliché 236 for what constitutes scientific progress: the notion that the theory which supersedes an older theory is truer. But once again we see that such a conception will not stand the test of historical reality. In this respect it is irrelevant that the Cartesian laws of impact evince certain immanent contradictions (which it might also be possible to eliminate). For the point *here* is that in any case it cannot be said, as would follow from the above-mentioned cliché, that the progress achieved by Huygens was based on the *empirical falsification* of Descartes's theories or on the *detection of new facts*. The first of these is not the case because, as we have shown, Descartes's assertions do not admit of such empirical falsifications and in fact are concerned with something entirely different from the assertions made by Huygens. Why the second point does not suffice can be brought to light in the following manner:

First: One of the most important reasons why Huygens and his successors moved away from Descartes lay in the rejection of Descartes's decision to view only those statements as scientifically demonstrated which were *clare et distincte* when considered in the light of reason. Thus Huygens expressly declared that he did not agree with this "$\kappa\rho\iota\tau\dot{\eta}\rho\iota o\nu$ veri" of Descartes.[15] Contrary to this he pointed out over and over again that his, that is, Huygens's, laws of impact were in "complete agreement with experience" while Descartes's "were contrary to experience."[16] Hence the first thing to undergo change here is what I have called a *judicative precept* (cf. chapter 4), a decision concerning the criterion governing the acceptance or rejection of a theoretical statement. In the present case, we are thus dealing with a transition in progress at that time from a strictly 237 rationalistic to a more empirical orientation, even though the latter should not be confused—at least not in the case of Huygens—with a strict empiricism. But this shift had already become so obvious and so second nature to Huygens's mind that he completely failed to see that his criticism and revision pertained more to *this judicative decision* than to Descartes's laws of impact as such. The reason for this is that Descartes's laws of impact, when viewed in the light of his "κριτήριον veri," are indeed empirically irrefutable and actually mean something quite different than in classical physics.

Second: We find a similar situation here regarding what I have called *normative decisions* for the sciences (cf. chapter 4), that is, decisions concerning the goals or ends which guide the sciences. As we have seen, Descartes had two such goals: on the one hand he wished to uncover the divine principles of construction, which he believed could only be revealed by reason; on the other hand he wished to make a contribution to the movement leading to the practical-technological use of such knowledge. He did not succeed in bringing about the harmonic unification of these two. Now, his critics, and those who came after him, have not always come out on the side of the latter goal, at least not with the same clarity that we find *expressis verbis* in their decisions concerning *judicative precepts.* But we can nevertheless infer this turn not only from the whole context belonging to the period following Descartes, but quite evidently as well from a consideration of the decisions made at the time concerning these judicative precepts, decisions which are in fact quite difficult to ground without a certain more or less conscious positing of goals. This follows since the demands for empirical proof simultaneously imply the need for controls in the form of verifiable and falsifiable predictions—and in this we already find an unmistakable element of praxis, namely, the ability to adapt oneself to anticipated results and to utilize knowledge of the future. Thus the moment Huygens laid down the relative reference system within which his laws of impact became valid, he had already given an indication of his intention *to give these laws a form in which they could be empirically realized via experiments.* This he considered to be so self-evident that Descartes became almost unintelligible to him, since he had paid little heed to this goal. Huygens did not see at all that Descartes could have had something completely different in mind. The traditional goal of investigating the divine ground and cause of the world, to which Descartes still held fast, even if he gave it a new, which is to say, mechanistic format, had not then really disappeared; however, it had been overlaid to such an extent by this other goal that it was partially obscured, at least in that it was no longer clearly distinguished from the latter.

Third: It was then only the altered judicative and normative conditions that led to the formulation of new *axioms* concerning laws of conservation; from these axioms in turn we finally come to the derivation of the particular laws of impact still in use today—*laws which are correct within this framework.*

On the basis of these three observations, it would seem to me that Mouy, the classical representative of the standard interpretation of the relation between Descartes and Huygens, errs when he summarily asserts:

"It is of special note that, in order to attain this remarkable result, Huygens uses Cartesian hypotheses as his point of departure. . . . Huygens is exclusively Cartesian in his principle, and it is impossible to find a more authentic example of the development of Cartesian postulates."[17] In opposition to this, however, when we view Huygens in terms of the concepts 239 presented in the previous chapter, we see that he did not merely *explicate* Descartes—that is, in a more correct manner than Descartes himself had done—rather, he *mutated* the Cartesian system, and this precisely in that he criticized the fundamental principles on the basis of which Descartes had proceeded. By way of proof for his thesis, Mouy gives five hypotheses underlying Huygens's theories, all of which he holds to be thoroughly Cartesian and from which all of Huygens's laws of impact are supposedly derived. But even if we overlook the fact that two of the five hypotheses deviate from their Cartesian counterparts—something which Mouy himself notes—we still find one, namely, the third as Mouy gives them, that can only be considered Cartesian when viewed in a rather superficial sense, that is, the "*principe du mouvement relativ.*" It is precisely here that we find one of the most important differences between Descartes and Huygens: For Descartes, this relativity is only something that exists "*en nostre pensée*"; thus it is excluded from physics, and consequently his laws of impact are developed from the point of view of a divine observer and based on the revelation of a deified and self-sufficient reason. Huygens, on the other hand, in a thought experiment proceeds on the basis of an expressly relativistic "earthly" reference system, one which thus admits of empirical realization. This view is linked to the image of a boat which sails along beside an even shoreline at a constant speed. (A corresponding picture adorns the title page of Huygens's treatise on motion.)

Therefore it is not the discovery of new facts which compels Huygens to distance himself from Descartes; rather, we are dealing here with a change in judicative and normative precepts. It is then only in the wake 240 of this mutation in the foundations of the system, and thus within the new framework which this mutation effects, that new axioms of physics are drafted and new methods of proof developed. Hence it is this mutation which opens up the way to a new perspective, to a new form within which questions are posed and answers obtained, or, in short, to a new kind of knowledge. *Thus the discovery of new facts is subsequent to the changes within the scientific-theoretical categories developed in chapter 4 and not the other way around: These changes do not follow the discovery of new facts.*

But from this there arises the question as to why these mutations in the cited categories occurred at all. It appears to me that the answer to this lies in the previously mentioned inconsistency within the Cartesian system, a system which awakened, in a fascinating and suggestive manner, the expectation of a practical-empirical mastery of nature, only to disappoint this in the end. Inextricably caught up within a scholastic-theo-logical framework of thought (*The Meditations* shows this all too clearly), Descartes was unable to hold fast to his normative goal. He could not satisfactorily carry off his project of formulating laws of impact, laying these down as the basis for a new physics in which they were then also supposed to be of the greatest possible practical use, while at the same time removing these laws again from all empirical involvement and val-uation, so that they might be formulated and projected only in accordance with a quasi-divine point of view, for which human time measurement and determination of motion are not applicable. Thus, in the Cartesian system, theology and the combination of reason, divine revelation, and mechanics all appear as a *bloc erratique*.[b] Pascal saw this quite clearly when he remarked that Descartes only needed God to give the world a shove, after which he was of no further use to him.

241　　Therefore we see that Huygens's advance over Descartes consists in the fact that he freed the Cartesian system of its fundamental internal contradictions, and not in the fact that he empirically refuted or improved the system on the basis of new forms of empirical knowledge. But this liberation was only to be had at the price of the mutation of the system. However, at the same time this mutation only came about through the use of all those elements of Cartesian philosophy which could be salvaged in the transition to the new framework, where they were interpreted anew and thus reappeared, as it were, in a new light. In particular, we find in this new system all those elements indicative of the fundamental ideas of a mechanistic point of view: hence, the reduction of all material occur-rences to action and counteraction, the principle of inertia as initially developed by Descartes, the Euclidean notion of space, etc. The new system was built up, so to speak, out of the remnants left over from the crumbling edifice of the Cartesian system. The solution to the problems posed by the inconsistency Descartes himself could not eliminate was achieved on the basis of the very same means that he had already made available and which survived the demise of his system.

In chapter 8, I depicted the flow of scientific-historical processes as a self-movement of system-ensembles. As I have already remarked, this is in fact only a kind of metaphor or image. But in what follows I will constantly make use of this expression as a kind of abbreviated formula for whenever I wish to point out that the development of science is not brought about by absolute, system-free factual assertions or absolutely valid first principles, but rather by the attempt to eliminate inconsistencies

b. Hübner uses the German *ein erratischer Block*. However, this is taken from the French *bloc erratique* given here. The term applies to a geological phenomenon of the Ice Age in which stones are pushed great distances by glaciers, as well as being broken up and strewn about: thus the literal translation "wandering stone." From this comes a more figurative meaning, "foundling" (in German *Findling*).

　　　　　　　　　　　　　　　　　　　　　　　　　　　　　　　CHAPTER 9

and instabilities within given historical systems, an attempt which always involves the retention of certain elements pertaining to the systems in question. The transition from Descartes to Huygens offers us an example of this. In point of fact this example shows that it is not new experiences that bring about new theories, but new theories, built up out of the residue of older theories, that offer a new horizon for experience, thus creating, as it were, new conditions of the possibility of experience. Moreover, this transition also illustrates the further thesis presented in chapter 8, that thesis which asserts that in the attempt to eliminate inconsistencies and instabilities within given systems (the self-movement of a system-ensemble), one part of the comprehensive framework of systems is accommodated to another part, with which it is not in agreement, and that it is precisely this process which constitutes the kind of progress which produces the mutation of judicative, normative, and axiomatic foundations of a system. In the present case, for example, this means that Descartes's laws of impact were brought into agreement with those of his normative principles which advanced the demands of the empirical-practical side of the system, or at least its experimental applicability.

Finally this chapter should also have shown how useful general scientific-theoretical considerations (in this case, those concerning the notion of progress) can be for the interpretation of individual historical processes. In terms of the present example, we see that the misconception which, as I understand it, underlies all previous interpretations of the transition from Descartes to Huygens has its roots in a lack of understanding of the scientific-theoretical categories properly suited to the explanation of this process. This is then the reason why it was so difficult to escape the constantly repeated clichés which offered themselves so readily and as such precluded further reflection.

10

THE SIGNIFICANCE OF THE HISTORICO-GENETIC VIEW FOR RELATIVISTIC COSMOLOGY AND THE CLASSICAL QUESTION OF WHETHER THE UNIVERSE IS AN IDEA

243 In this chapter we will be involved in a further working out of the discussion and role of the a priori which was begun in chapter 8. This we will do by relating it to the example of relativistic cosmology. Initially we will not deal with the question of the justification of the a priori. This question of justification will become the subject matter of a later section of the chapter. With the justification of the a priori, the relation of relativistic cosmology to reality will also be clarified. We will then be able to answer, in a new and appropriate manner, the old question raised by Kant: Is the universe merely an idea?

Our reason for singling out relativistic cosmology from the numerous cosmologies on hand today has nothing to do with its particular characteristics. Everything that will be demonstrated with regard to this cosmology could also be shown with respect to the others. But this choice does prove advantageous when addressing a larger circle than that composed solely of specialists. For this reason I have also generally avoided a detailed discussion of the latest developments in the area of cosmology, such as those pertaining to the work of Hawking, Penrose, Wheeler, and others. These developments call for significant technical expertise, which cannot be generally assumed.

Relativistic cosmology is based on the *general theory of relativity* and
244 usually on the *postulate of the cosmic substratum*[a] and the *cosmological principle* as well. Let us begin by examining each of these in turn.

a. In German this reads *Postulat des Weltsubstrates*. Here this has been translated as the "postulate of the cosmic substratum." The precise meaning of "cosmic substratum" is given in section 10.2, p. 141. For the moment it is enough to note that this obviously refers to a kind of material base permeating the universe as a whole and conceived as if it were a fluid or gas, hence "cosmic fluid," which appears in some of the contemporary literature on the subject.

10.1 Einstein's A Priori Foundation
for the General Theory of Relativity

When Einstein formulated the general theory of relativity, he did not initially give any thought to the possibility of describing the physical facts at hand more exactly or of detecting any new ones. As described in chapter 8, Einstein was concerned above all with gaining a more *unified* and *illuminating* picture of nature by means of a different *interpretation* of these facts. In this vein he had already been successful with the special theory of relativity. There it had been his intention to eliminate the contradiction between Maxwellian light theory and the classical principle of the equivalence of all inertial reference frames. But the special theory of relativity, which had in fact eliminated this contradiction, could not on the other hand be unified with the theory of gravity. This unification was first achieved with the general theory of relativity, thus with the introduction of Reimannian geometry into physics, which allows *all* coordinate systems, not merely those of inertial reference frames, to be considered as equivalent, by identifying the paths of all freely moving test bodies with geodesics, regardless of whether the movements are classically viewed as determined by inertia or are seen as being determined by gravity. It was only at this pont, then, that Einstein achieved what he had been seeking: a conceptual framework for the development of a comprehensive theory unifying Maxwellian theory, mechanics, and the theory of gravity.

At first it was by no means certain that Einstein's theory would offer *empirical* advantages over the other theories which had been developed in connection with the same areas of experience. But Einstein saw the superiority of his theory precisely in that it was more comprehensive than the others. *In this* lay its real justification, indeed the only justification that was initially available. In the formulation of this theory Einstein was guided by the fundamental principle that *nature is determined by a unitary system of connections* (*einen einheitlichen Zusammenhang*). And this fundamental principle is a priori insofar as it can never be falsified. The failure of every theory developed with the help of this principle would thus be explainable precisely because these theories had not hit upon *that particular* unity which lies at the basis of nature. Hence we could also call this fundamental principle a regulative principle in Kant's sense, since it simply states that we *ought* to look for the unity of nature.

In the meantime we have come to recognize that the general theory of relativity does present us with empirical advantages over Newtonian gravitational theory. Does this then mean that Einstein's original arguments have become superfluous? Not at all, as is evident from the previous chapters. There, the following was established: An empirical confirmation states almost nothing about the content (*Inhalt*)[b] of a theory, thus about

b. In chapter 11, Hübner will distinguish between the *Inhalt* and the *Gehalt* of a theory (pp. 155–57). Both terms translate into English as "content," and it is quite difficult to

the truth or falsity of what is being asserted by its axioms. This is to say that we can only confirm the basic statements derived from these axioms; and since, according to the rules of logic, true statements can follow from false ones, we may now state that from such confirmation we can only conclude that nature has not expressly said no to the content of the theory—but neither has it said yes to it. Hence this content always requires a priori foundations and justifications *above and beyond* such confirmations. These are essential even in the event that the theory is later viewed as having been falsified. For in that case we know only that nature has rejected *something* pertaining to the many assertions contained in the theory, but we do not know precisely what has been rejected (on the Duhem-Quinean problem, see chapter 4, section 4.1). Confronted with this, one is again compelled to give a priori and other reasons for what is to be retained and what rejected.

246

But let us then read what Einstein himself has to say. In his opinion, the positivists—and today we can add the Popperians—hold a "primitive or elementary ideal" when they see as the only task of science the making of empirically correct predictions.[1] He explicitly thinks it to be entirely possible "that there are any number of possible systems of theoretical physics all with an equal amount to be said for them."[2] Accordingly, reasons quite different from empirical reasons would have to be presented as a basis for choosing among these systems. In the end, the "investigator's passion"[3] is primarily aimed at making "reality" understandable; thus, insofar as it is directed toward the *content* of the theory, this passion is directed at something about which observational data cannot, for the above-mentioned reasons, provide sufficient information.[4] *For this, in his eyes the most important, aspect of a theory, Einstein gives as his criterion precisely the regulative principle mentioned above, a principle which is in complete accord with the postulate of the harmonization of a system-ensemble developed in chapter 8. He writes: "We are seeking the simplest possible system of thought which will bind together the observable facts."*[5]

247

"The special aim which I have constantly kept before me," he states, "is logical unification in the field of physics."[6] "In a certain sense, therefore, I hold it true that pure thought can grasp reality, as the ancients dreamed."[7] The relation with experience obviously remains intact; but the theoretical construction which is thrown over this has its own additional context pertaining to its foundation and justification, one which is independent of experience. And this context obviously belongs to the *self-movement of the system-ensemble* as described in chapter 8. In addition we read that

distinguish them except by context. In general we might say that *Inhalt* carries with it the notion of "meaning"—everything that goes into making a theory true or false, etc. *Gehalt,* as used in chapter 11, relates rather specifically to Popper's formal notion of *Wahrheits- und Falschheitsgehalt* (truth and falsity content), and as such to a "certain amount" of truth or falsity contained in a theory.

"the structure of the system is the work of reason."[8] This was "perfectly evident" to him even at a time when he still believed that he had to proceed on the basis of the assumption that "we can point to two essentially different principles"—namely, the general theory of relativity and the Newtonian theory—"both of which correspond with experience to a large extent."[9] This amounts to saying that neither of these two theories could claim any essential *empirical* advantage.

10.2 The Postulate of the
Cosmic Substratum and the Cosmological Principle

These two postulates (the so-called cosmological principle is in fact also a postulate) are usually introduced to provide a transition from the general theory of relativity to a relativistic cosmology. 248

The postulate of the cosmic substratum states that the universe is to be conceived as a fluid with regularly distributed material density, the molecules of which, for instance, are galaxy clusters. These molecules, which participate in the currents of this fluid, should be motionless with respect to their immediate surroundings. All coordinate systems or observational positions should always be thought of as firmly connected to the cosmic substratum.

The cosmological principle states that the universe presents the same view to every observer. In terms of classical physics, this means more precisely that at corresponding points of equal coordinates within different coordinate systems, matter has the same speed, the same momentum, and the same density. Within the framework of the general theory of relativity, this means (roughly speaking) that the geometrical relations within the universe are the same for every observer of comoving referent frames.[10] In order to be able to consider these relations as isotropic and homogeneous, the world-lines of the galaxy clusters must extend radially to and from the center of the coordinate system of the observer.[11]

It is immediately apparent that these two fundamental principles are also characterized by the notion of the uniformity, or what we might better call here the simplicity, of nature. Once again we must note that this notion, as in the case of the general theory of relativity, is not developed in abstraction, but rather in relation to a given body of physics—hence in relation to a given historical situation. Both the postulate of the cosmic substratum and the cosmological principle find their origins within the context of the conceptual framework coined by classical mechanics. Furthermore, the cosmological principle is formulated here within the geometricization of physics. By means of such all-embracing principles, there arises a picture of the universe as a physical unity. The postulate and the principle in question here draw their persuasive power from the notion of the simplicity and uniformity of nature. This persuasive power is no substitute for an empirical confirmation of relativistic cosmology, nor 249 could it be simply invalidated by an empirical falsification; for as we have already seen, such a confirmation would state too little about the content of the postulate or the principle, and it would be impossible to ascertain whether a falsification referred precisely to them.

10.3 Four Possible Cosmological Models of
Relativistic Cosmology and Their A Priori Discussion

All of the presuppositions necessary to the formulation of relativistic cosmology are now on hand. From the postulate of the cosmic substratum, as well as from the cosmological principle, the form of the time-dependent
250 metric of the universe follows logically as expressed by the so-called *Robertson-Walker line-element*.[12]

If you insert the metric tensors given by means of this line-element into the field equations of the general relativity theory, then a cosmological formula for the universe can be derived in the end which admits of several possible solutions, thus allowing for several possible interpretations of the history of the universe.[13] Here I select only four *types* of solutions, and hence four types of cosmological models.[c] It would be superfluous to examine all of the types, since the same philosophical problems that occur in relation to these four also occur in relation to all the rest. In the following discussion, then, we will only be concerned with these problems. The four types of solutions are the following:

251
1. The universe has existed for an infinite time and is spatially finite. During this time it either has remained completely unchanged or has expanded (Einstein's model).
2. The universe has existed for a finite time. In the beginning everything was compressed together into one point. Then, following a kind of primordial explosion, or "big bang," it began a constant, nonreversible expansion.
3. The universe exploded at a finite point in time long ago in a "big bang"; but when it has reached a certain maximum expansion, it will begin to contract again.
4. Originally the universe was infinitely extended, and hence its material density was infinitely low. Gradually it contracted; but when it reached a maximum density, it began a renewed expansion to infinity.

Again we can show that there are significant reasons, independent of empirical verification, that speak both for and against each of these cosmological models. Here we will merely present these reasons, without

c. It is obvious from the context that these "cosmological models" (*Weltmodelle*) are models for the history of the universe, that is, temporal models concerning both the past and future, of the universe. For the meaning of "cosmological" in this context, especially as regards a regressive series, see Kant, *Critique of Pure Reason*, "System of Cosmological Ideas," B 435ff.

going into a more detailed examination of them, except in a few instances. How any decisions can be made in such cases will be treated at the end of this chapter.

In what follows, Kant's First Antinomy will quite clearly be called to mind. In fact, this will come to the surface several times in the course of the discussion. However, the a priori discussion of the universe in terms of relativistic cosmology will show that we can no longer maintain Kant's claim that this antinomy involves a necessary dialectic of reason. Therefore whether or not the universe is an idea is no longer a question that   252
can be resolved today within the framework of Kantian philosophy.

Let us consider a problem which concerns all four cosmological models, namely, the problem that all these models obviously presuppose a universal cosmic time. In point of fact the cosmological principle already implies such a notion. For if the geometrical relations in the universe gradually change in the same manner and in all directions for everybody alike, then this means that they all change at the *same time*. But such simultaneity, such universal cosmic time, is only possible for *selected observers*—namely, for those who are not moving (either in an accelerated or in a decelerated manner) relative to the mean density distribution of their surrounding matter—and this obviously means for those observers who are moving along with the cosmic substratum. Here the fundamental relativistic principle of the equivalence of all reference frames loses its meaning and significance.

Now, viewed in this light, relativistic cosmology does not contradict the laws of the general relativity theory since it is in fact always possible to select from among the generally admitted group of observers those who observe, in a *certain respect* and under *particular conditions,* the same thing. But the question must nevertheless be raised regarding what right we have to speak of a *universal cosmic time* that can only hold for certain reference frames. And this question is obviously not one that can be answered by means of some kind of experiment.

Two opposing views resulted regarding this, one that defended universal cosmic time and one that rejected it.

Eddington and Jeans, for instance, spoke out in favor of universal cosmic time, employing reasons that were again aimed at eliminating the schism between the scientific and the prescientific view of time that had   253
been generated by the general relativity theory.[14]

In contrast, Gödel turned against a notion of universal cosmic time, as this pertains to the physical universe, precisely because it "depends on the particular way in which matter and its motions are arranged in the world" and remarked that "a philosophical view leading to such consequences can hardly be considered as satisfactory."[15]

Here we have two opposing opinions with two philosophical foundations. The dispute can thus only be resolved philosophically.

Let us now discuss each of the cosmological models of relativistic cosmology in terms of those kinds of criticism or justification of the content which are independent of empirical examinations.

The *first model* mentions an *infinite time* for which the universe has existed. Here again we find opposing opinions: One opinion holds such an infinite time to be a priori possible; the other asserts that such a time is a priori impossible.

Kant rejected infinite time on *logical grounds*. An infinite time that has been running its course up to the present is in itself a contradiction, since an *end* point, the Now, would thereby be posited for something *in*-finite, while according to its concept the infinite can have no such end point.[16]

However, it obviously escaped Kant's notice that this contradiction only arose because he had already presupposed a particular concept regarding the *existence of a whole*. According to his conception, such an existence can only be asserted if it is fundamentally possible to present *all* parts of a whole or, as he puts it, to complete the "synthesis" of these parts. Such a completed synthesis is in fact at variance with the concept of an infinite whole. In truth, then, Kant's method of proof is far from being logical as he assumed; it is rather epistemological (*erkenntnistheoretisch*), that is, based on the *reality* of an object (*Sache*).

This becomes even clearer when we recall Cantor's concept of the infinite. In the present context, Cantor can be regarded as Kant's opponent. Cantor holds that the existence of an infinite whole is given whenever there is, for instance, an enumerative procedure employing finite means to assign each part of this whole a particular number belonging to the consecutive cardinal numbers. Thus all that is required is that it be fundamentally possible to carry out this procedure with respect to *every single* part; but it is not necessary that *all* parts be presented. It is also not contradictory within this view to speak of an infinite whole that has a final, concluding member. The infinite series of negative numbers ending with -1 is an example of this.[17]

Kant proceeds then, as it were, with an *extensional,* Cantor with an *intensional,* concept of an existing whole. But it cannot be asserted that the intensional concept is *logically* impossible. The difference, as already stated, is entirely epistemological, hence philosophical. But one cannot avoid confronting this problem if the *content* of the first cosmological model is to be taken at all seriously, by way of either justification or criticism.

Before setting out the a priori arguments for and against the *second cosmological model,* we must point out that there are two possible interpretations of this model: Either you actually carry out the extrapolation of the curve representing this model to the point at which all matter in the universe was in fact compressed together or you view this point as a singularity, that is, as something bracketed-off from other occurrences in the universe.

The proponent of the first interpretation would have to admit that the result arrived at in this way on the basis of the existing body of physics contradicts this physics; for example, it runs counter to the laws of conservation of quantum physics. Thus this physics would have to be false according to the law of logic (the proposition "¬A follows from A" can only be asserted if A is false).

The second interpretation does indeed avoid logical difficulties of this kind; but then here we must take into account that the course of the universe in a finite time is not physically definable with respect to the entire span of time. Thus physics would be unable to say anything about the beginning of the universe.

The disputed point which presents itself *here* [with respect to the second interpretation] can be formulated in the following question: Do we or do we not wish to concede validity to a cosmological model that compels us to make this assumption? Regardless of the decision, we again see that it is not an empirical question; rather it depends upon the *normative* expectations we set for a physical theory, that is, what kinds of results we demand from it. And of course this expectation and this demand are for their part dependent upon the clarification of that question concerning our right to view nature as a set of interrelations that can be exhaustively interpreted by means of physics.

256

A further philosophical question that can be raised in conjunction with the second cosmological model has to do with the notion of *finite universal cosmic time:* Is such a time a priori possible or not?

Here again we have a classical statement by Kant: If the universe has only existed for a finite period of time, then an empty time must have preceded it. But in an empty time nothing could have begun; for in it no one part could be distinguished from any other—and Kant obviously means by this that an earlier part could not be distinguished from a later one.[18] St. Augustine had already noted, however, that the assumption that the universe had a beginning *before* a particular point in time in the past in no way implies that it had a beginning *in* time.[19] This amounts to saying that the universe could have come into being coincidentally with the time belonging to it.

Thus Kant's objection no longer applies here. But another objection was raised regarding Augustine: If we accept the above-mentioned view, then the beginning of the universe would be an event without precedents and would therefore be "objectively" impossible, since the "objectivity" of events has to do with the fact that they can be ordered in a continuous causal interconnection.[20]

The acceptance of this view once again depends upon whether or not we ascribe, along with Kant, a kind of "transcendental meaning" to the causal principle.

Now let us consider the status of the *spatial finiteness of the universe,* something which is asserted by numerous cosmological models.

Here again Kant's argument obviously proves to be unconvincing. In a treatment parallel to that of finite time, he asserts that a notion of a finite world must be connected with that of an empty space within which this world would be located, and then utilizes this to show the absurdity of such a notion since an empty space would in fact be "nothing."[21] However, the argument fails because the finite universe of relativistic cosmology need not be embedded in a space surrounding it, and hence is not necessarily located in an infinite, flat space.

But it would nevertheless be too simplistic to push Kant aside merely by remarking that the only geometry he knew was Euclidean. For the question of whether Euclidean geometry is preferable to, or has a transcendental meaning over and above, other geometries remains unsettled; and neither the demonstration of the existence of non-Euclidean geometries nor the relativity theory has resolved this question once and for all. Kantians and operationists influenced by Dingler still maintain today that non-Euclidean geometries are purely mathematical, fictitious creations that have nothing to do with the reality of the space of the universe. The Kantians found their opinion on a certain theory of intuition, the operationists on a certain theory of measurement. And while these theories can be disputed, they certainly cannot be refuted, for the reasons mentioned earlier, by appealing to the *empirical* successes of the relativity theory. The arguments of the Kantians concerning the role of intuition and those of the operationists concerning the role of measurement can only be addressed within the respective realms of the theory of intuition and the theory of measurement. In the discussion of the fourth cosmological model we will return to this point, as well as to Kant's view that empty space, like empty time, is nothing.

The *third cosmological model* forces us to ask whether the described change from expansion to contraction implies an oscillation which is *cyclical*. Might this be possible, say in the sense of Nietzsche's eternal return of the same, or would it be impossible?

If the course of time is determined by means of a series of consecutive states of the universe, then the return of the same state of the universe would also mean the return of the same point in time. But then there would be absolutely nothing—not even a temporal differentiation—by means of which the original state could be distinguished from a later state. That which is perfectly identical in this sense could not return at all.

The notion of the eternal return of the same could be salvaged by introducing an absolute time that is independent of the states of the universe; but since relativistic cosmology does not allow for such a time (this must not be confused with the universal cosmic time which hinges upon selected observers), this path is barred to it.

But the cycle of oscillation implicit in the third cosmological model need not be viewed as the return of exactly the same state; rather, it can be viewed as the return of something similar. This can be asserted because then the oscillation involved would merely be a function of the universal cosmic time of selected observers, and it follows that the same states would likewise exist only for *these* observers. Furthermore, the requisite homogeneity of the immediate surroundings would only be an approximation, even for these observers. Hence it would indeed be entirely possible to interpret the third cosmological model in terms of cycles, but we are not absolutely compelled to do so.

By contrast it would also be conceivable that the universe had a beginning in which all matter was densely compressed into a small area and that after a period of great expansion it will once again end up just as it began. But if this view is accepted, then two questions have to be explained, as we have already mentioned, namely: Can there be a *first* event? And can there be a *final* event? For it is just as difficult to relate the former to a causal interconnection as it is the latter. 259

Let us now turn finally to the *fourth cosmological model*. This posits an infinite, empty space, both at the beginning and at the end of the universe.

The reason that Kant rejects both empty space and empty time lies in his assertion that neither could be the object of intuition.[22] The weakness in this argument can already be seen in the fact that at another place in the *Critique of Pure Reason* Kant states precisely the opposite. There he says, "We can never represent to ourselves the absence of space, though we can quite well think it as empty of objects."[23] And regarding time he writes: "We cannot, in respect of appearances in general, remove time itself, though we can quite well think time as void of appearances."[24]

The principal question here is whether the assertion that neither empty space nor empty time is representable says anything at all about their existence. Machean empiricism, which understood representability as "possible experience," was convinced that the question had to be answered in the affirmative. Thereby Machean empiricism aided in the formation of the general relativity theory and in the attempt to refute the notion of absolute space by means of the principle of the equivalence of all reference systems, since absolute space obviously cannot be an object of possible experience. 260

With regard to this, however, it is especially important to note that under certain conditions something like absolute space nonetheless results once again from the Einsteinian field equations. Above all it can be shown that in the case of a vacuum, that is, in the absence of matter, the curvature of space-time does not disappear. Thus space also has a thoroughgoing structure, even where there is nothing in it—it has an existence for itself. De Sitter, for example, demonstrated that the cosmological field equations also allow for a solution in the case of an empty space. If a test body is introduced into such a vacuum, it moves in accordance with the inner structure of empty, and hence absolute, space.[25]

The way in which various physicists responded to this is quite interesting.

One group, among them Dicke, attempted to alter the general theory of relativity such that this contradiction to the empiricist philosophy standing behind the a priori postulate of the equivalence of all reference systems could again be overcome.[26] On the other hand, Synge was not at all put off by the reemergence of absolute space in the very lap, as it were, of the general theory of relativity, as long as this followed consistently from the theory and had no direct effect on its unified foundational context (*einheitlicher Begründungszusammenhang*).[27] In this case, Dicke and Synge are merely defending a priori fundamental principles; for here it is not at all a problem of *empirical* difficulties encountered by the general theory of relativity, but rather purely a matter of certain *mathematical* consequences which follow from the theory and which allow for statements about space that appear to be either a priori acceptable or a priori unacceptable. It is not a matter of making decisions *from* experience, but rather *for* experience; that is, it concerns the theoretical *presuppositions* for the interpretation and description of reality.

With regard to this, we can say the same thing for Newton as well. He did not prove the existence of absolute space experimentally, even though he believed he had. In his opinion, the existence of absolute space evidenced itself in that bodies accelerating relative to space showed certain effects, for instance, centrifugal forces. But at least since the formulation of the general theory of relativity, we have known that centrifugal forces relative to absolute space only constitute one possible *schema of interpretation,* a schema which lies at the basis of classical physics.

Having recounted the most important a priori grounds that must inevitably be discussed in connection with the foundation and evaluation of the *content* of relativistic cosmology, we must now ascertain how matters stand with respect to the empirical examination of this theory. To this end, I have selected one of the most interesting cases to serve as a characteristic example.

10.4 The Difficulties Involved in Falsifying Relativistic Cosmology
At this point we are presented with another occasion for contesting the philosophy of Popper. Popper holds that the most important aspect of a theory is its ability to be falsified and subjected to empirical testing.

Accordingly, for him everything else involved in and leading up to the formulation of theories belongs to the realms of psychology, the history of science, or metaphysical belief, and as such has nothing to do with a scientific *foundation*.

In *The Logic of Scientific Discovery,* Popper writes: "The act of conceiving or inventing a theory, seems to me neither to call for logical analysis nor to be susceptible of it."[28] Indeed, concerning scientific discovery he goes on to write: "And looking at the matter from the psychological angle, I am inclined to think that scientific discovery is impossible without faith in ideas which are of a purely speculative kind, and sometimes even quite hazy; a faith which is completely unwarranted from the point of view of science, and which, to that extent, is 'metaphysical.'

"Yet having issued all these warnings, I still take it to be the first task of the logic of knowledge to put forward a concept of empirical science, in order to make linguistic usage, now somewhat uncertain, as definite as possible, and in order to draw a clear line of demarcation between science and metaphysical ideas—even though these ideas may have furthered the advance of science throughout its history."[29]

In opposition to this view, our previous considerations have shown the exceedingly important and decisive role played by a priori reasons, reasons not based on experience, precisely for the *scientific justification of the content* of a theory—in the present case, for the justification of the content of the theory of relativistic cosmology. But there is yet another aspect to be considered here. Beyond this, the following section should show that the processes coming *at the end* of a scientific investigation, and as such meant to serve as empirical tests for it, cannot simply be separated, as Popper believes, from the a priori considerations which serve to found the investigation *at the outset*. If one fails to provide foundations, then so does the other. Thus what leads up to the formulation of a theory is no more "scientifically non-discussable"[d] than what belongs to the testing of a theory: it is just as little a matter of mere faith, no more unclear, and it is not merely of psychological or historicogenetic significance. To the contrary, it constitutes one of the most important objects for the "logic of knowledge."

From relativistic cosmology an equation can be derived that determines the dependence of the observable energy radiated from a galaxy upon the redshift of the light emitted from it. This equation admits of three types of solutions, depending on whether the curvature parameter has a value of -1, 0, or $+1$.[30] We would then be able to speak of a falsification of relativistic cosmology if the curve resulting from the measured data pertaining to the radiation energy and the corresponding redshift showed itself to be incompatible with all three types of solutions.

263

264

d. Cf. addendum to note 29, this chapter.

Indeed, it is still not possible to collect the data necessary for such a falsification, owing to the present state of telescope technology; but this has no essential significance here. Rather what is significant is the following: The above-mentioned equation gives us the testable relation between the observed radiation energy and the redshift only if we introduce a postulate which asserts either that galaxies emit the same amount of radiation at all times, or at least that for all galaxies this radiation is dependent upon time in the same manner.[31] Such a postulate, however, is nothing other than a specific instance of the cosmological principle, according to which conditions in the universe are supposed to be the same everywhere.

Let us now assume that telescope technology had advanced to a point where the necessary data could be gathered, and further that we attained a curve by means of which relativistic cosmology was falsified. We could then dispute this falsification by abandoning the postulate which made the falsification possible in the first place. Admittedly, along with this we would have to accept that the theory would no longer be falsifiable in the above-mentioned manner. But if we nevertheless decided to accept the falsification, then by retaining the postulate in question we would be retaining precisely that aspect of the theoretical presuppositional framework for which we have the fewest intelligible reasons. For it would constitute something of an overextension of the cosmological principle if we employed a belief in the homogeneity of the universe to underwrite our acceptance of the notion that all galaxies should exhibit such uniform behavior. But regardless of what we might decide to do in such a case, it is clear that a decision about the falsification, indeed about the very falsifiability, of the theory depends in every case upon how we treat the general principle that has already played a decisive role in the formulation of the theory. Hence we cannot simply separate the "beginning" of a theory from its "end": each is inextricably tied to the other. *Quod erat demonstrandum.* Thus Popper's strict criterion for the demarcation of the scientific and the nonscientific cannot be maintained, since this criterion rests upon the assumption that such a separation between that which is empirically falsifiable and that which is not empirically falsifiable is possible.

10.5 Concerning the Justification of the A Priori in Relativistic Cosmology

Let us summarize what we have found thus far: Relativistic cosmology rests upon the a priori assumption of the unity and simplicity of nature. Judgments regarding the possible cosmological models based on this theory must be made in part on empirical grounds and in part on a priori grounds. But while in the first case the a priori element, namely, the assumption just mentioned, also plays a decisive role, in the second, even subsequent empirical testing cannot simply replace or weaken the a priori judgment already made concerning the content of the theory, because such testing does not yield sufficient information about this content.

Thus, if we indicate, as Popper does, that a priori grounds are only of psychological interest and do not admit of scientific discussion, then not only relativistic cosmology but also the entire body of physics is lost. For indeed these grounds can be shown to exist everywhere in physics, being only particularly evident in the case of relativistic cosmology.

If we now ask about the justification of the a priori fundamental principles of relativistic cosmology, then we find once again, in direct contradiction to Popper, that the answer is only to be obtained with the help of the historicogenetic aspect of science, which he so grossly undervalued. Thus here again we come to recognize quite clearly how Progress I and Progress II come into play in the manner indicated in chapter 8 and how it is that the context for justification arises within a given system-ensemble.

As we have now seen, the historical situation present for Einstein can be described as follows: *Given* was the incompatibility of the special theory of relativity and Newton's gravitational theory. Also *given* was the postulate of the simplicity of nature. Presupposing all this, it was necessary for Einstein to bring the special theory of relativity and the gravitational theory into agreement. This he accomplished by means of the a priori postulate of the equivalence of all reference frames, which led to the general theory of relativity.[32] *In this* lay its justification, a justification relative to the given situation and in accord with the postulate of the harmonization of the given system-ensemble. Now a new situation was given. It consisted once again of the postulate of the simplicity of nature and now in addition of the general theory of relativity. From this there resulted the further necessity of unifying the general theory of relativity and existing cosmology. The postulate of the cosmic substratum and the cosmological principle result from this, as well as from ideas underlying classical mechanics. And thus these too have an a priori justification which becomes intelligible in terms of the situation and which cannot be replaced by any sort of empirical data. If in addition we wish to see the justification for the postulate of the simplicity of nature, which in fact runs through everything here like a red line, then we would have to reach much farther back to the spiritual-intellectual upheavals that led to the dissolution of nominalism and the Christian-mythical world view. For within this earlier context the universe was seen as anything but uniform and simple. (It was neither homogeneous nor isotropic: it had an "above" and a "below," a "left" and a "right"—the heavens, the earth, and hell.)

The various arguments that speak in an a priori manner both for and against the individual cosmological models derivable from relativistic cosmology are also subject to being judged only in terms of a particular situation, that is, in terms of a *situational logic* as described in chapter 8. For the most part here we must agree with Kant in his rejection of all possible cosmological models when these are considered in relation to Newton's a priori space-time philosophy and the particular context for philosophical problems in which Kant found himself. He was only wrong in maintaining that the a priori foundations of Newtonian philosophy and the problem context created primarily by this philosophy were *necessary* and thereby given by reason for all times. When we cross over to relativistic cosmology, we find a completely altered situation. All of those a priori arguments which speak *for* each of the four cosmological models, and which from the very first exclude none of these models, are now justified on the basis of the a priori postulates of (1) the general theory of relativity (the equivalence of all reference systems), (2) the cosmic substratum, and (3) the cosmological principle. For now the critique of the finiteness of the universe, insofar as this rests upon the exclusive validity of Euclidean geometry, can finally be refuted by referring to the new a priori right to assume relativistically curved space as well as Euclidean space. A point in favor of the intensional definition of infinite sets within this situational context is that it fits harmoniously into relativistic cosmology whereas the extensional definition does not. Now everything favors a philosophical conception that limits the significance of intuition and causality for cosmology in that it assesses this significance in terms of the *entire a priori contextual framework* and *justificational context* out of which relativistic cosmology arises and which it depicts.

269

Obviously, the justifications under discussion here have the weakness of again pertaining to a particular situation, and *only* to this situation. Furthermore, they do not even necessarily result from this situation, as would be the case in a strictly logical inference. Other possibilities for reacting to a situation in an a priori manner are in no way excluded. At any particular time several theories are usually in the running, competing, as it were, against one another. Hence, aside from relativistic cosmology, there are numerous other modern cosmologies, such as the new version of Newtonian cosmology, the steady-state theory, the cosmologies of Eddington, Dirac, Jordan, etc. And in the end even the a priori structures (*Apriorismen*) lose their justification when the historical context to which they are bound is replaced by a completely different one. Thus, as a rule, they can never be considered as *binding* (*für zwingend*); certainly they can never be thought of as universally necessary.

But as we have also shown here, there is a middle ground between absolute validity on the one hand and a complete, "scientifically nondiscussable arbitrariness" or mere "faith" on the other; namely, there is a rational, comprehensible, and intersubjectively understandable (*intersubjektiv nachvollziehbares*), a priori form of argumentation regarding a particular historical situation. Here we must rid ourselves of the misguided and illusory notion that the a priori is identical with that which is eternally valid, the transcendental. We must recognize that even the a priori is caught up in the movement of that which is historical, while at the same time recognizing that this does not make it something empirical.

Obviously empirical arguments also have a part to play in this movement. But arching over everything that results from observation as a confirmation or falsification, there are powerful a priori constructions, essentially independent of observation, which are continually shifting, changing, and taking on new form, and which exert and bring about their own, often revolutionary, effects. These considerations will now allow us to answer the question of whether the universe is merely an idea.

10.6 Is the Universe Merely an Idea?

Kant based his conviction that the universe is only an idea on the following argument:

1. The universe *either* has a beginning in time and is contained within spatial limits, *or* this is not the case and it "is infinite as regards both time and space."[33]
2. It is *logically* false that the universe is infinite; and it is *epistemologically* (*erkenntnistheoretisch*) false that the universe is finite.
3. Since the statement about the universe in (1) presents a complete disjunction and both sides of this disjunction are false, it follows that no true statement can be made about the universe. The universe is thus only an idea in that there is no reality existing in itself which corresponds to its concept; and this is proved by means of an unavoidable dialectic of reason.

Now, as we have already mentioned, this dialectic only occurs if the argument is based, among other things, on Newton's a priori space-time philosophy. If we take relativistic cosmology as a basis, then the absolute disjunction in the first premise of Kant's argument disappears. Furthermore, we have shown that the four cosmological models cannot be refuted *logically,* but rather only philosophically. Moreover, they can also be the object of *empirical* testing. Finally, there is probably no one who is still willing to assert that Newtonian physics is a *necessary* result of *reason.*

Thus Kant's dialectic collapses. On the other hand, our example of relativistic cosmology shows that even today we must treat the universe as an idea. But this does not come about because every cosmological model that can be developed for it is false by necessity, but rather because every model of this kind is only an a priori construction, concerning the content of which there can never be sufficient empirical evidence. Such constructions are therefore neither false nor true in themselves; and they only admit of comparison with previous or subsequent creations to the extent that the situations out of which these creations arise are themselves comparable.[34] Nevertheless they have their empirical and above all their historical–a priori foundation and justification *within the limits just mentioned*. For this reason, as ideas they are then also part of the world in which we live, the world we find historically present at a particular time, and a part of our own self-understanding in terms of this world. This we can assert because the manner in which the universe *appears* to us within a particular situation with *virtual* necessity is itself of great significance for this self-understanding. For this reason the idea of the universe is a part of *our* reality. If the objection is then raised that this reality is nonetheless only a shadow cast by us, then we might reply that as *our* shadow it is a shadow we cannot jump over. And finally, what would we be without a shadow? A ghost.[35]

272

The last two chapters were intended to provide deeper insight into the observations of chapter 8, concerned with the structure of historical processes, by illustrating what was said there in terms of several pertinent examples drawn from the history of science. But now we must once again examine the question of *truth* in greater detail. Indeed, we have already seen that there can be little talk of a continual approach toward something like an absolute truth—an old dream of science. But Popper's philosophy might still give the impression that an application of the concept of truth that is thought to be logically unobjectionable would have to be incompatible with a historistic view, such as the one presented here. We will now address ourselves to this error.

11

THE CRITIQUE OF THE CONCEPT OF TRUTH IN POPPERIAN PHILOSOPHY; THE CONCEPT OF TRUTH IN THE HISTORISTIC THEORY OF THE EMPIRICAL SCIENCES

Popper bases his philosophy on Tarski's theory of truth.[1] However, he alters Tarski's formulation "P is true if and only if p" so that it reads "P corresponds to the facts if and only if p." Here P stands for a linguistic expression, p for the corresponding fact. From Tarski, Popper also takes the three minimal conditions that must be met if such a truth and correspondence assertion is to be free of contradiction and possible: First, it must be formulated in a metalanguage in which it is possible to speak about the expression belonging to the corresponding object-language, for instance P. Second, it must be possible to describe in the metalanguage all the facts that can be described in the object-language, for instance p. Third, the metalanguage must contain expressions like "correspond to the facts." In turn, it then also becomes possible to define what a real fact is, namely, "p is a real fact if and only if P is true."

Now Popper thinks that we can never know with certainty if something is true. But he nevertheless believes that with the aid of Tarski's theory of truth we are justified in speaking about an *approximation* to the truth.

It is in this vein that Popper initially defines the content (*Gehalt*) of a statement a. This consists in the class composed of all logical inferences which can be drawn from a. Hence, if a is true, then this class only contains true statements. But if a is false, then this class contains both true and false statements. (For example: "It always rains on Sunday" is false, whereas the inference that it rained last Sunday can be true.) By the truth content (*Wahrheitsgehalt*) of a, Popper correspondingly understands the class of logical inferences drawn from a that are true statements; and by falsity content (*Falschheitsgehalt*), the class of logical inferences drawn from a that are false statements.

Now let us assume that two theories, T_1 and T_2, are comparable. In such a case Popper remarks that T_2 is closer to the truth, or corresponds better to the facts, than T_1, if and only if

a) the truth content, but not the falsity content, of T_2 exceeds that of T_1;
b) the falsity content, but not the truth content, of T_1 exceeds that of T_2.

155

If we then let $Ct_T(a)$ stand for a conceivable measure of the truth content of a and $Ct_F(a)$ for a corresponding measure of its falsity content, then we can give a formula for the verisimilitude of a, $Vs(a)$, which reads as follows:

$$Vs(a) =_{Df} Ct_T(a) - Ct_F(a) .$$

(Later on Popper improved on this definition; but we need not go into this here.)

A theory would have a maximal verisimilitude if all facts to which it corresponded were true; in such a case it would be absolutely true according to Popper. In actuality, however, we can only speak of a greater or lesser degree of verisimilitude and correspondingly only of an approximation to the truth. Popper calls objective truth, absolute truth, and verisimilitude regulative ideas.

11.1 Critique of Popper's Metaphysical Realism;
the Concept of Truth Belonging to the Historistic Theory of Science

It is rather remarkable that Popper and his advocates could earnestly believe that purely logical analyses and definitions, like the ones just presented, could support a "metaphysical realism." For logic can make no assertions about reality, whereas metaphysical realism consists nonetheless in the assertion that there *exists in itself* a reality which transcends consciousness—precisely that reality which we are supposed to come to know it terms of an ever-increasing degree of approximation. Indeed, Tarski and Popper have shown that there can be no *logical objection* raised against the concepts of truth and verisimilitude—but this is all they have shown.

It can be readily seen that their exact definitions are compatible with theories of knowledge other than metaphysical realism. For example, consider the theory which is the polar opposite of metaphysical realism, namely, the metaphysical idealism of Berkeley. Since Berkeley proceeds on the basis of the principle "esse est percipi," or "being is what is perceived," a fact is then only that which is perceived by a subject and nothing more than this. Thus, using Berkeley, we could replace "fact" with "what is perceived" and correspondingly arrive at the formulation: "P corresponds to what is perceived if and only if p." This would also satisfy all of Tarski's minimal demands for statements of truth, though such a formulation might appear somewhat strange at first glance. But here it is only necessary to point out clearly that for Berkeley there is nothing except that which is perceived. Thus who could doubt that, when viewed from a purely logical perspective, Berkeley's "esse est percipi," along with his notions of facts, truth, etc., derivable from this, is to be considered possible—no matter what other stance might be taken concerning his theory?

Thus Tarski's and Popper's definition of truth only receives epistemological *content and meaning (erkenntnistheoretischem Inhalt)* when what is meant by "fact" and "corresponds to a fact" is clarified. However, as has been shown clearly, especially in chapters 5 and 10, Popper interprets this concept in the sense of a naive empiricism and realism.

If on the other hand we proceed on the basis of the position developed in the course of our own arguments, then "facts" are things which arise via perception *and* historically conditioned a priori structures. We can abbreviate this as "interpreted perception." Additionally, the expression "corresponds to facts" always implies a prior positing of certain judicative conditions as well (cf. chapter 4). Thus, seen in this light, the proposition "P corresponds to the facts if and only if p" would explicitly mean "P corresponds to the interpreted perception under such and such judicative presuppositions if and only if p."

If we then understand the a priori structures of a particular perceptual interpretation in accordance with chapter 8 as a system of principles S, then the phrase "corresponds to the facts" is obviously only possible in relation to a given S. In other words, "is true" means "is true in S." Thus, when explicitly formulated, "P is true if and only if p" reads "P is true in S if and only if p." Here again we find complete agreement with Tarski's purely logical prerequisites, since what is demanded here is indeed only that the semantic metalanguage also be a meta*theoretical* language, and thus that it contain not only such expressions as "is true" and "corresponds to the facts," but the expression "is true in S" as well.

However, not only is the Popper and Tarski concept of truth compatible with theories of knowledge other than that of metaphysical realism, but this holds as well for Popper's definition of "approximation to the truth." This definition rests on the truth and/or falsity content of a statement a, as well as on the comparability (*Vergleichbarkeit*)[a] of theories. Hence, if, as the present context demands, we understand the chosen measure for the truth content, $Ct_T(a)$, to mean more explicitly $Ct_T(a)$ in S, and that of the falsity content, $Ct_F(a)$, to mean more explicitly $Ct_F(a)$ in S, then it follows from the Popperian definition that we will correspondingly have to speak of a verisimilitude in S such that $Vs(a)$ in $S = {}_{Df}(Ct_T(a) - Ct_F(a))$ in S. This means that we assert a more or less greater degree of verisimilitude to a under the conditions given in S. Thus, just as there can be no objection to speaking of truth itself when it is only relative to a par-

a. Here "comparability" (*Vergleichbarkeit*) must be understood quite literally as a condition belonging to the theories by virtue of which they *admit of* or *allow for comparison*. These latter expressions will sometimes be used as a substitute.

ticular S, so too there can be no objection to speaking of an approximation to the truth under the same conditions.

The concept of explication or Progress I detailed in chapter 8 offers us the simplest example of this. It consists (among other things) in the fact that the truth content of a theory is recognizably augmented when an increasing number of true statements can be derived from it—whether it be through more exact determination of the constants of the theory or improved mathematical methods for attaining confirmable prognoses, etc. We can say here, then, that the verisimilitude in S becomes increasingly greater. And the same holds when we are dealing with theories that *admit of comparison* owing to an overlapping of respective realms, each of which at the same time belongs to an S as a subset.

278

An elementary instance of this occurs when theory T_1 is based on S, theory T_2 is based on S', and S is a subset of S'. In such a case we might then find that Vs(a') in S' is greater than Vs(a) in S; and we could thus speak of a greater approximation to the truth in S' than in S. But this alters nothing as regards the necessary relation to an S.

As previously stated, the concept of the approximation to the truth is only possible for theories insofar as they admit of comparison, and thus only in relation to an S to which they both belong in an overlapping manner. The systematically founded and historically exhibited multiplicity of mutually exclusive, incommensurable S's or of S's which bear a mere family resemblance to one another, which was brought to light in the previous chapters, is so constituted as to preclude any talk of an approximation to *the* truth by means of *any and all* of the sciences. Such an approximation can only be spoken of in relation to a *particular* S.

Thus we can summarize: Neither the concept of truth nor that of the approximation to the truth is necessarily connected with the concept of a metaphysical realism or that of an absolute truth. To the contrary, here again it is the theory of historical a priori precepts and system-ensembles that proves to be thoroughly compatible with both of the above-mentioned concepts.

Correspondingly, it now also becomes possible to speak of a regulative idea within the framework of this theory, namely, that idea which would continually draw us on to carry out an ever-increasing number of explications of a theory, to seek continually for more and better confirmations of it, and thus to attempt to ensure that it is better than another *comparable* theory. Admittedly, even if we were to succeed in reaching a maximal verisimilitude such that in the end our theory corresponded to all possible facts relating to it (whatever this might mean), we would still be unable to assert with certainty that this theory was consequently absolutely true, precisely because it would always be logically conceivable that the same thing could be achieved to the same extent via another comparable theory. We could then imagine, for instance, two or perhaps more completely

279

satisfactory interpretations of the universe. This is logically conceivable because the total set of facts confirmed by one theory is not identical with the total set of facts confirmed by another, and, in addition, because the basic statements utilized in the proof of theories are also variously interpretable within the frameworks of different theories. A brief simile might serve to illustrate this point: Let us assume that we could give a perfectly adequate description of the world by putting on either a red or a blue pair of glasses, but in no other way. In the first instance everything would be red, while in the second everything would be blue. How could we know that one of the two offered us the absolutely true picture of reality?

Popper also gives us a simile: He describes absolute truth as a mountain peak that is always enshrouded by clouds. Even supposing we were able to reach it, we still would not be able to know that we had actually done so. But the absolute truth would still be there as something objective, remaining unaffected by whether or not we were able to know it. Thus, taking up the earlier example again, it would be possible for everything to be "really" blue, without our ever being able to decide that this is the case. To this we might reply that an absolute truth of this kind would be totally meaningless for us. On the other hand, however, it is of the *greatest meaning and significance* for us to know truth and verisimilitude *in an S;* and this we can do in a *well-founded* and *objective* manner. For example, let us consider the following statement: "If I presuppose the microphysical theory T, together with certain instrumental and judicative precepts, then there is an electron cloud at this location." This statement can be known to be empirically true or false with all the exactness and certainty that could be desired; and this truth or falsity can be known objectively, which is to say that it can be known by anyone. However, the statement does not relate to an absolute object, but rather to an object which is only given via the *relation* to a set of principles. Hence absolute truth and the approximation to absolute truth are neither knowable nor of any meaning. On the other hand, a truth and an approximation to truth are both knowable and meaningful in terms of a given S and can also be objectively founded as such.

Here again it becomes clear that the Popperians confuse a *"relational concept of truth"* (related to an S) with a "relativistic" concept of truth, a concept which corrupts all knowlege by reducing it to the level of capriciousness, arbitrariness, and mere assertion. Moreover, the relational concept of truth has nothing in common with skepticism and agnosticism. Contrary to this, it brings to light the conditions under which objective truth, though not a conditionless or absolute truth, is alone possible. Who would dispute that we see "true things," though the whole world knows that our observations depend upon the particular constitution of our eyes and optical nerves? The things are "true" within the framework of the conditions of the apparatus of sight and learned sight habits.

Popper distinguishes four different theories of truth: The *correspondence theory,* the theory which he himself embraces (truth of a statement
as correspondence to the facts); the *coherence theory,* which is concerned
only with freedom from internal logical contradiction and the logical interrelations of a system; the *evidential theory,* where knowledge of truth
and objective truth are one and the same; and the *pragmatic theory,* where
truth is a matter of utility. He holds all theories except the correspondence
theory to be subjective; and by this he means that they fail to do justice
to the necessary relation to objectivity implicit in the concept of truth.
The coherence theory fails to take into account the relation to a reality
that transcends consciousness; the evidential theory overlooks the fact
that truth occurs first and foremost in itself and only then is discovered;
the pragmatic theory fails to recognize that practical results tell us nothing
about the truth of a matter. Over and against this view, we can now state
the following about the theory which I have proposed here: *First,* it is a
correspondence theory insofar as it requires correspondence to the facts,
albeit in an S. *Second,* it is a coherence theory to the extent that it involves
an expectation that the given S will contribute to the harmonization of a
given system-ensemble, and thus that this S, as the condition of possible
empirical truth and a possible approximation to truth, will serve to free
this system-ensemble of ambiguities or contradictions and to disclose the
comprehensive structure of the interrelations belonging to it. *Third,* it is
an evidential theory insofar as it requires that truth be fundamentally
knowable, and thereby demands a reason for speaking of truth at all. And
fourth and finally, it is a pragmatic theory of truth in that it allows us to
speak about only that truth which stands in a known or knowable relation
to a given S within a given system-ensemble, and is thus a part of the
world within which the people of a given time practically move, act, and
objectively experience reality.

Finally, the Popperians maintain that only the regulative idea of absolute
truth makes possible and encourages scientific progress. As far as I can
see, this is the real reason that they were led to this idea in the first place.
To this I counterpose here the regulative idea of a truth in relation to an
S; furthermore, the regulative idea of the approximation to truth in relation
to S (Progress I); and finally, the regulative idea of an optimal formulation
of this S as the condition of possible experience within the framework of
a given system-ensemble (Progress II). In comparison to *this* idea, Popper's goal for progress shows itself to be pretentious and counterproductive. To use his simile again, the desire to climb to a peak so situated
that you can never know whether you have actually reached its summit
is more frustrating than encouraging. On the other hand, progress becomes
a meaningful idea when the goals set are such that one could know
with certainty whether they had been reached. Accordingly, not only is
the idea of an absolute truth meaningless, but it is now equally evident that

the only reason the Popperians give in support of it is anything but cogent. We cannot expect the light in which we see things to be the light of the things themselves; but we do have the right to assume that nature *appears* to us in this light, that it becomes *visible* in this light, and that it shows itself to us as it *actually looks in this light*. Whoever would have more than this demands of himself that he be a god or that he have divine revelation. An absolute truth would have to be based on the thing in itself. To connect an empirical theory with this, as the Popperians attempt to do, is something which lacks all intelligible foundation.

11.2 Concerning the Truth of the Historistic Theory of Science Itself

At this point we must now consider a question which is certainly of   283
decisive importance here, namely: What is the status of the truth of the historistic theory of science itself? Is this theory only true as well within an S as a part of a historical system-ensemble, and thus is it itself subject to the fate of a possible future mutation? If this were the case, then the theory would not be contradiction free. For it would seem that in order for the theory to remain valid, it would itself have to be caught up in the self-movement of a system-ensemble. Hence its doctrine would be confirmed precisely in its being annulled and superseded (by a mutation). Or is it then absolutely true? But how could such a presumption be justified? Why should this theory be the sole exception to a history in which every other theory has arisen only to fall again?

Let us look back once more and recall the particular sections which constitute the foundation of the theory. In chapter 3 we found on the basis of a purely logical analysis that basic statements, theorems (natural laws), and axioms pertaining to an empirical theory have a priori precepts as a necessary condition. In chapter 4 these precepts were systematized, again on the basis of a logical analysis, and arranged in categories. Further, it requires no experience to see that these precepts can be neither logically, transcendentally, nor arbitrarily founded and that consequently only one possibility remains, namely, to found them within a historical context. In chapter 8 we then found—and this will be dealt with more extensively in chapter 13—that such a historical context, should it ever become an object for *science,* is only describable in terms of axiomatic theories, and *there-*   284
fore is only describable as a system-ensemble. Finally, in chapter 8 it was also shown, again logically, that the manner in which a founded development (explication) and change (mutation) come about and the manner in which these occurrences run their course within such a scientifically constructed system-ensemble can only be understood in terms of the concept of the harmonization of system-ensembles. *At the end of all these logical reflections, which in their entirety result from working out the concept of empirical science in terms of axiomatic theories, we arrive at*

*the logical conclusion that history, when scientifically viewed, must be
construed as a kind of self-movement of system-ensembles.*

The empirical examples drawn from the history of science and brought
into play in the previous chapters serve merely to illustrate this point;
they do not constitute a foundation for the historistic theory of science.
Aside from their "intrinsic" scientific-historical value, these examples
have a propaedeutic meaning in the sense indicated in chapter 4.

Hence the historistic theory of science is neither empirically nor transcendentally true; rather it asserts a *logical truth* like that belonging to
an if-then proposition. This might be briefly formulated as follows (cf.
chapter 9, pp. 136–37): *If* there is an empirical science, *then* this science
must, when concerned with history, view it either explicitly or implicitly,
as a history of self-moving system-ensembles. This logical truth *as such*
is always valid; hence it is not itself historical—thus the contradiction
mentioned at the beginning of this section does not pertain to it. But on
the other hand, it is nevertheless tied to a historical condition insofar as
the object to which it refers, that is, science itself, might disappear at
285   some future time. Admittedly, the theory would retain its validity in such
a case, but it would quite literally forfeit its actuality—it would be obsolete.

However, if science were to disappear by giving way to another manner
of viewing and considering things, then this occurrence would obviously
still be an occurrence in history from the point of view of this new way
of considering matters; but it would no longer be describable in the system-
historical manner typical of science. In chapter 15, for example, we will
consider what history can mean in terms of a mythical interpretation.

11.3 A Few Additional Critical Comments
concerning the Most Recent Form of Popperianism

By way of concluding we might bring in a few additional comments concerning the critique of Popperianism.

I should like to single out two major theses which are presently representative of the philosophers at the London School of Economics (LSE).[2]
These can be stated as follows:

1. Falsificationism—which is simply another word for "Popperianism"—
   is superior to inductionism because it is thoroughly deductive and
   avoids Goodman's paradox. (What this entails will be explained later.)
286   2. No fact that is used in the construction of a theory can confirm this
   theory.

Let us begin with the first thesis. This is founded in the following
manner: The LSE philosophers maintain that genuine scientific inferences
are deducible and arise in one of two ways: either we draw testable logical
inferences from premises containing conjectural suppositions; or we begin

with a tested and falsified logical inference, and move to the denial of a set of premises from which the falsified inference was deduced.

To this we can reply that, as we have shown, every falsification also has particular premises, for example, axioms pertaining to particular theories of observation. Now, if these premises are also conjectural suppositions—and the LSE philosophers do not deny this—then the falsification is as well only a conjectural supposition. This supposition can be purely arbitrary, in which case the falsification is rendered practically meaningless; or the scientist will have particular *reasons* for these suppositions—but then it is absolutely impossible to avoid inductions. If, for example, you employ a ruler for a measurement, you would not normally expect the ruler to have changed since the last time you used it. But this expectation, which cannot be avoided if the measurement is to have any meaning, will be dependent for the most part upon previous experiences, and is therefore inductive. Hence, what advantage has actually been gained over the inductionists? The falsificationists have reproached them by saying that they permit verifications which are nothing more than conjectural suppositions; but the inductionists might just as easily turn the tables and with equal justification reproach the falsificationists for sanctioning falsifications which are also nothing but conjectural suppositions, or what is even worse, for sanctioning falsifications which are completely meaningless.

The second reason the LSE philosophers give in support of their view that falsificationism is superior to inductionism consists in the assumption that Goodman's paradox only pertains to inductionism. This paradox runs as follows: First, the word "gred" is defined in the following manner: Something is "gred" either if a test performed at time t shows it to be green, or, if it is not tested at time t, it is red. Next the statement "All emeralds are green" is compared with the statement "All emeralds are gred." Thereby we find that paradoxically both statements are inductively confirmed to the same extent, since it is obvious that every inductive confirmation of the first is also an inductive confirmation of the second. However, it is not true that this paradox pertains only to inductionism and not to falsificationism. If we investigate an emerald and if the result of our investigation at time t_0 reads "The emerald is not green," then the hypothesis "All emeralds are gred" would be falsified. If, however, the result at time t_0 reads "The emerald *is* green," then the falsification fails, and according to Popper the hypothesis would be corroborated. The LSE philosophers obviously fail to see that Goodman's paradox does not relate principally to the problem of inductionism, but rather to the problem of how a *genuine* law can be distinguished from a *pseudo*law. Thus Goodman's paradox shows clearly that it is not sufficient to define a law, as Popper does, as "a statement that must be falsifiable." For even though "All emeralds are gred" is a statement that can in fact be falsified, no one would seriously consider it to be a genuine law. Accordingly Goodman's paradox pertains to inductionism and falsificationism alike.

In addition it should be noted that the strict rejection of inductionism is incompatible with the general rule of the Popperians already dealt with in chapter 5, namely, the rule stating that one theory or research program is better and more progressive than another with which it is competing if it is confirmed by more facts than the other. For this rule can itself be nothing other than a rule of induction. Indeed, it points up that the kind of program thought to be more progressive, to offer more for the future, and thus meriting more support, is judged to be so on *inductive* grounds, namely, because it has *hitherto* proved to be particularly fruitful.

One last word regarding this matter. The criterion for progress just mentioned proves to be completely inadequate to the task of interpreting transitions like the one between Descartes and Huygens analyzed in chapter 9. There we found that the transition was not primarily a matter concerning theories about facts; rather it revolved first and foremost about the problem of working out a normative criterion for the determination of what a scientific theory ought to be in the first place. It was a transition from a purely rationalistic mechanics to an empirical mechanics. Hence, in terms of this example, the criterion of the Popperians can only become valid once the empirical mechanics has been introduced and is on hand. The Popperian philosophy cannot of itself produce an interpretation of those preeminently important scientific-historical processes which involve the working out of metatheories and scientific-theoretical systems within which we first come to grips with the problem of whether and to what extent facts ought to play a role, or what is and is not to be accounted a fact in the first place.

Let me then summarize: First, in terms of that very rule (mentioned above) which they themselves introduce, the Popperians contradict their own radical anti-inductionism. Second, this rule is itself ambiguous (What is a fact?) Third, this rule fails to take into account one of the most important areas of progress, namely, the progress which pertains to the movement from one metatheory to another (a kind of progress which on the other hand, as the previous chapters have shown, can be interpreted by means of the theory of historical system-ensembles.)

Now let us look briefly at the second thesis of the LSE philosophers. Here by way of example they refer to the attempt to improve the Newtonian theory by bringing it into line with the perihelial movement of Mercury, which is correctly predicted by the general theory of relativity. According to them, this attempt is to be rejected because the knowledge of the fact of this movement is used in the construction of the improved version of the Newtonian theory. The following can be said in opposition to this: Let us assume that at some time in the far distant future, when the history of science has been completely forgotten, someone uncovers two books: one containing the improved Newtonian theory; the other, Einstein's theory. There is no doubt that this person will consider both theories to be completely equivalent, since they both correspond to the facts to the same degree. This shows that the expression "using a fact in the construction of a theory" is misleading. It appears to point up a kind of logical circularity. But in truth the person who prefers the relativity theory to the improved Newtonian theory, because Einstein formulated his theory without trying to adjust it to the movement of Mercury, does not base his preference in any sense at all on the criterion "supported by a fact," but rather on something of a completely different nature, namely, on sheer priority in time: It was Einstein who first derived the questionable perihelial movement, not Newton. However, we cannot find a justification of greater scientific value in such an argument, since in the final analysis both theories are in agreement with reality. This alone is decisive. Over 290 and against this, the history of the production of this agreement is without meaning. Further, we might ask: What gives the believers in absolute truth, like the LSE philosophers, the right to assert that one theory, adjusted so as to conform to the known facts, is falser or simply less true than another theory merely because the former did not succeed all at once, like the latter one?

12

CRITIQUE OF THE SNEED-STEGMÜLLER
THEORY OF SCIENTIFIC-HISTORICAL
PROCESSES AND SCIENTIFIC PROGRESS

Sneed and Stegmüller base their work on the idea that a theory can be described in terms of its mathematical structure by utilizing the definition of a set-theoretic predicate (*eines mengentheoretischen Prädikates*).[1]

To illustrate this let us take the example of classical particle mechanics (CPM). Here we can say:

CPM(x) $\longleftrightarrow$ there is a set P, a set T, an $\bar{s}$, an m, and an $\vec{f}$ (translated: x is a CPM if and only if there is a set P, etc.), such that the following hold:

1. x = $\langle P, T, \bar{s}, m, \vec{f} \rangle$ (translated: x is a structure with P, T, etc., where P denotes a set of particles, T a set of time points, $\bar{s}$ the function of the position vector, m the function of mass, and $\vec{f}$ the function of force).
2. P is a finite, nonempty set.
3. T is an interval of real numbers.
4. $\bar{s}$ is the function of the position vector with $D_I(\bar{s}) = P \times T$ and $D_{II}(\bar{s}) \subseteq \mathbb{R}^3$. ($D_I$ is the definition domain of $\bar{s}$; "$\times$" is the Cartesian product. Or in other words: In the domain of $\bar{s}$, a particle is in every case coordinated with a time point. D_{II} is the range of $\bar{s}$, thus the image set in which the domain is mapped by the function. $\mathbb{R}^3$ denotes the set of the triples of real numbers. Thus "$D_{II} \subseteq \mathbb{R}^3$" means: The range (that is, the set of position vectors) is a subset of the set of the triples of real numbers—since every position vector is defined in terms of three real numbers, i.e. its coordinates.)
5. m is a function with $D_I(m) = P$ and $D_{II}(m) \subseteq P$, where $m(u_i) > 0$ for all $u \in P$. ("$u \in P$" means that u is an element of the set P.)
6. $\vec{f}$ is a function with $D_I(\vec{f}) = P \times T \times \mathbb{N}$ (where $\mathbb{N}$ denotes the set of natural numbers in which the number of forces operating on a particle is mapped) and $D_{II}(\vec{f}) \subseteq \mathbb{R}^3$. Further, for all $u \in P$ and $t \in T$, $\sum_{i \in \mathbb{N}} \vec{f}(u,t,i)$ is absolutely convergent (i.e. the sum of the absolute values has a limit).
7. For all $u \in P$ and $t \in T$, $m(u) \times D^2\bar{s}\,(u,t) = \sum_{i \in \mathbb{N}} \vec{f}\,(u,t,i)$ (where D^2 denotes the second time derivative of $\bar{s}$; thus here we are concerned with the well-known equation mass $\times$ acceleration = force.)

Now, according to Sneed and Stegmüller, this purely set-theoretic definition enables us to make an empirical assertion which might be expressed thus: There is an application a of this structure to real systems. For example, the solar system is supposed to have such a structure. According to this view, empirical statements of this kind are taken to have the following general form: a has a structure that is defined by a particular theory—abbreviated, *a has an S,* where S is called the fundamental law of the theory (e.g. the CPM).

In a second step, Sneed and Stegmüller define the meaning of a *theoretical quantity.* Theoretical quantities, they maintain, are those which are given via a theory-dependent measurement. According to their conception this means that the determination of the values of these quantities is dependent upon a fruitful previous application of precisely those theories in which the quantities in question occur. Thus, for example, only force and mass are theory-dependent in the CPM, while position and time are not, since these can also be the object of nonmechanical forms of measurement—for instance, optical measurement.

Every application a of a theory is called a *model of S* and is distinguished from a *possible (potential) partial model* of CPM. Thus, for instance, particle kinematics (PK) is a possible partial model of CPM. In terms of the previous definition: $PK(x) \longleftrightarrow$ there is a set P, a set T, and a function š, all of which fulfill the above-mentioned points 1–4. Mass and force are then left out or abstracted (and this obviously applies to point 1 as well, where the terms initially occur). In this way CPM appears as a "theoretical expansion" of PK. So instead of generally saying, as before, that "a is S," we can now say the following (using "(I)" to distinguish this from a later formulation):

(I): a is a possible partial model of S; and there is a theoretical expansion x of a which is a model of S. This is called the "primitive form of the Ramsey formulation of the empirical content of a theory."

But why should the rather complicated statement (I) take the place of the simpler "a is an S"? This is justified by Sneed and Stegmüller in the following manner: In order to check "a is S," you must determine the values of certain theoretical quantities. However, according to the definition of these quantities, this means that a successful or fruitful application of the theory with structure S must be presupposed. In order to check this application, another previously successful application would again have to be presupposed, etc. The result would thus be an infinite regress or a circle. So in order to avoid this and to bring out the empirical truth of (I), Sneed and Stegmüller argue that it is sufficient to know whether the *nontheoretical* quantities employed in the description of a satisfy (I). For example, the confirmation of (I), interpreted within the framework of the PK and/or CPM, rests solely on the supposed demon-

293

294

stration that there is some particle to which a time interval and a position vector can be said to correspond. For then there would be a theoretical expansion which is a model of the CPM—which is to say, we can apply the theoretical functions of the CPM to such an *a,* which presents us with a possible partial model of the CPM. Thus, in comparison to "*a* is an S," (I) is an attenuated empirical assertion, namely, the assertion of a merely possible, not an actual, application of the theory-dependent quantities.

12.1 Critique of the Sneed-Stegmüller Definition of Theoretical Quantities

At this point certain critical questions already become apparent, especially concerning the unsatisfactory nature of the Sneed-Stegmüller definition of theoretical quantities. Why must the successful application of that theory upon which such quantities are dependent enter into the definition? Does not the conclusion which Sneed and Stegmüller draw from this here—i.e. that space and time are not theory-dependent quantities—show precisely how questionable this is? For as we have seen on numerous occasions in the previous chapters, there are no space or time determinations which do not involve complicated theoretical presuppositions. Indeed, in order to be able to assert that "*a* is an S" is true, measurements must be employed; but this can only occur if the validity of the theories necessary for these measurements has at least in part been posited a priori (this too has been demonstrated on numerous occasions). We cannot ask for more than this because it leads to the circularity or infinite regress pointed to by Sneed and Stegmüller themselves; but in fact we *need* not ask for more than this. Furthermore, even the assertion "*a* is a possible partial model of S," if, as claimed, it is to belong to the empirical content of the theory, is such that it can only be judged by means of measurements. Accordingly, as soon as the Sneed-Stegmüller standpoint is accepted, how can the formulation of the empirical content of a theory given in (I) escape the very difficulties already inherent in the statement "*a* is an S"? Finally: Is (I) really an adequate definition of the empirical content of a theory? Are we really able to maintain with absolute certainty that this statement cannot be given the meaning "*a* is a possible partial model," *which means* that it is *a priori interpretable* in this manner? *In this sense* the empirical content of a theory would then consist, for example, in statements like: "The planetary system, interpreted a priori within the framework of classical mechanics, exhibits such and such *particular* movements, masses, and forces," etc.

12.2 Critique of the Sneed-Stegmüller Distinction between the Core-Structure and the Expanded Core-Structure of a Theory

Let us now continue to follow the Sneed-Stegmüller theory in its further development. There we find the following: For every theory there are various "*intended applications*" (e.g. for classical physics, the solar system, the tides, the pendulum, etc.). These applications are held together by certain "constraints" ("*Nebenbedingungen*"). Thus the same functional values are allotted to the same object in different applications. (E.g. the mass of the Earth is taken to be the same within the framework of the solar system and within that of a subsystem of it.) Further, in certain applications "*special laws*" hold, by which are to be understood "*specializations of the structure S.*" (E.g. in terms of CPM we have the law of gravity, Hooke's law, and many others.) If we take all this into account as well, then according to Sneed and Stegmüller we arrive at the completed form of the Ramsey formulation of a theory. This runs as follows:

(II) There is a theoretical expansion χ of the set μ of physical systems to models of the mathematical structure S such that the theoretical functions employed in this expansion satisfy a class of previously given constraints and, in addition, such that certain genuine subsets of μ are expandable to models of certain more precise (specialized) formulations of the structure S.

In this, the so-called core-structure C of a theory is distinguished from the expanded core-structure E. To C there belong: (1) The set of possible models (i.e. the mathematical structure of the theory). (2) The set of possible partial models. (3) The restriction function (the function by which the possible models are coordinated with possible partial models). (4) The set of models. (5) The set of constraints. If we add the above-mentioned special laws to these five constitutive elements of the core-structure (which need not be discussed here in greater depth), then we arrive at the expanded core-structure E.

Now the central idea of the Sneed-Stegmüller theory is that C remains constant in the course of the scientific-historical process, while E changes. (II) is thus an empirical assertion which fluctuates with time. The core-structure, on the other hand, is, as it were, the a priori element in the theory; hence it remains constant and stands in need of no special "immunization strategy" ("*Immunisierungsstrategie*").

But then we are unavoidably confronted with the following critical question: Is not the line of demarcation between C and E arbitrary? Where are the criteria in terms of which this line can be objectively or necessarily defined? Moreover, if these criteria are lacking, then so too are the grounds for determining what can be empirically sacrificed and what cannot. Admittedly, the set-theoretic definition of a theory does allow the immune a priori core to become clearly visible in a formal manner. But this nevertheless presupposes that an intelligible foundation is already on hand that could explain why the boundary between the core-structure and its expansion lies where it does and has not been drawn otherwise. For example, in order to establish that the second law of classical mechanics—

force = mass × acceleration—belongs to the core-structure, and thus should be accounted a priori valid, it is necessary to turn to something completely different from set-theoretic inquiries. Stegmüller himself rec-
298 ognizes this when he mentions that the measurement of the quantities determined by this law already presupposes this law. Further, it is erroneous to suppose that the core-structure of a theory requires nothing in the way of immunization. As the previous chapters have shown, all a priori elements of a theory are exposed to historical erosion; and it is precisely for this reason that they must be constantly defended anew in various ways (partly by bringing in other a priori principles for their defense, and partly by demonstrating their ability to deliver a sturdy framework which might serve to support a fruitful and meaningful world of experience).

12.3 Critique of the Sneed-Stegmüller "Dynamics of Theories" (Theoriendynamik)

We now come to the definition of a theory for which yet another set J of physical systems is introduced, representing the "intended applications" of E. Further, the class of possible intended applications of E is now designated A(E). In this way we arrive at the following statement: x is a physical theory if and only if x = ⟨C,J⟩. Thus, according to Sneed and Stegmüller, we can also now express the empirical content of a theory in a set-theoretic manner. In place of (II) we now find:

(III) J is an element of A(E) (since μ in (II) is nothing but the set J of intended applications taken at a particular time and E is the more precise formulation of S in (II)).

Finally we also find a definition for what it means "to have a theory at your disposal at time t"—this being defined both semantically and pragmatically.

299   The *semantic definition* reads: A person P has a theory T at his disposal at time t if, at time t, P knows (1) that J is an element of A(E); (2) that this E is the most precise known formulation of that E to which the application J belongs; and (3) that J is a maximal set belonging to the application E. The *pragmatic definition* reads: A person P has a theory T at his disposal at time t if (1) T is at P's disposal in the semantic sense; (2) there exists another person P_0 (for example, the creator of the theory) who has established the intended applications of T by means of a paradigmatic set of examples J_0; (3) J_0 is a subset of the chosen set J for P at time t; (4) P believes that there is a more precise E' of E such that J is again an element of A(E'); and (5) P believes in an expansion of J such that this expansion is also an element of the more precisely formulated E'.

Point (4) of the pragmatic definition is accordingly called "theoretical belief in progress" while point (5) is called "empirical belief in progress."

With this we now have the means at our disposal to deal with the major issue here—namely, to develop the Sneed-Stegmüller theory of scientific-historical processes, which goes by the name of a "dynamics of theories."

The *historical* fact that different people often embrace the same theory, while nevertheless connecting different hypotheses with it, is only to be explained in that the people in question have differing opinions as to the expanded core-structure E or the scope of the set J of intended applications, but nonetheless hold fast to the same paradigmatic primary set J_0.

The *historical* fact that theories are so frequently falsified (here this means the falsification of statements (II) and/or (III)), without being abandoned, can be explained by pointing out that it is only the attempted *expansion* of the core-structure which fails, while the core-structure itself remains unaffected and in fact can never be overturned. (Stegmüller cites the following example: If light is not composed of particles, then this does not mean that particle mechanics as a whole fails, but only that one element of the set of intended applications belonging to its expansion E has been removed or canceled.)

300

Seen in this light, we find that a *normal* scientific development occurs when E and J are expanded, whereas a *revolutionary* development occurs when a new core-structure C is developed.[2] This then explains the further *historical* fact that revolutionary developments never find their point of origin in the falsification of some core-structure, as such falsifications are in fact quite impossible. Thus, despite the failure of various expansions of the core-structure, the core itself remains in use until a better one is found. If an object is urgently needed by man, then it is better to have it in a rather unsatisfactory form than not to have it at all. Moreover, the failure of various expansions of the core-structure is thought to be no proof that such expansions are absolutely impossible.

301

Especially when dealing with Stegmüller, one gets the distinct impression that the set-theoretic definition of a theory is thought to be the single most important element figuring into and allowing for the explanation of the three fundamental facts pertaining to the history of science described above. This would mean that if one were to consider a theory in a manner which was not set-theoretic—e.g. as used to be the case, as a class of statements (the statement conception) which can be true or false as a whole—then the immunity of the core of the theory, which is, so to speak, beyond "truth and falsity," would be unintelligible; accordingly, the course of the history of science would be reduced to an irrational enigma.

Now, as we have already pointed out, the set-theoretic presentation of a theory does in fact offer the advantage of exhibiting the a priori core of a theory in a formal and particularly clear manner. But this presentation does not in any way do away with the necessary decision preceding it concerning *what* ought properly to belong to the core of the theory, that is, to the a priori element of the theory, and what not; what criteria are to be employed for such a decision; and what is generally to be understood by the term "a priori" in such a context. Decisions of this sort can only be made, however, by checking the individual statements to find out if they can be founded, whether it be in an empirical or nonempirical manner; and in doing so, the entire complex within which these statements occur is drawn into the process. No *formal* presentation, like the set-theoretic one, can save us from investigations concerning content. Therefore this task of formal presentation will always belong to the *end* or final stage in the development of a theory. In other words, this task will only become possible once the theory exists as a class of statements which have been fully presented. It can never pertain to the beginning or initial stages of a theory, that is, to the period when the boundary line delimiting C and E, the line distinguishing the a priori and the a posteriori, is first established, when the axiomatic, functional, judicative, and normative precepts receive their own a priori justification and are thereby distinguished from the empirical givens which only come to light within the framework created by means of these precepts.

302

Because the set-theoretic description of theories hides the problematic nature of the foundation of these theories under the guise of its formal advantages, it also hides their historical conditioning. Thus we find that Stegmüller gives a purely psychological explanation for the continued adherence to a core-structure in the face of the failure of many attempted expansions pertaining to it, briefly stated again, the explanation that it is thought to be better to have something deficient than nothing at all. Within the context of the mode of analysis belonging to the set-theoretic description of theories, it becomes impossible to view this kind of adherence *historically* rather than psychologically; consequently we are cut off from the possibility of seeing this adherence as something which is objectively founded in a particular situation—a view which is corroborated by the many pertinent examples cited in the previous chapters and which finds its universal foundation in the theory of historical system-ensembles.

Finally, we also find that this set-theoretic description can never explain *why* revolutionary changes come about, *why* a new core-structure is suddenly developed. What Stegmüller has to say about this—namely, that such changes occur when the old theory becomes reducible to the new theory and when the new theory offers at least as much by way of explanation and prognosis as the older one—has nothing to do with the set-theoretic description of theories; rather it merely represents a purely historical assertion, which, as the previous chapters have also shown, is incorrect. However, when we consider scientific changes in terms of their system-theoretic description, and thus in terms of the historistic theory of science developed here, we come to understand why the historical development did not run its course according to this assertion, and why, despite this, the process did not become enigmatic and irrational.

303

Nonetheless, the Sneed-Stegmüller theory does prove, on the whole, to be a useful instrument for the analysis of given theories, even though it is questionable in certain respects. However, a "*dynamics* of theories" (Theorien*dynamik*), and by this we mean a metatheory of the origin, foundation, selection, and historical development of theories, cannot be derived from it; accordingly, this name seems to be an inappropriate and misleading designation for this theory.

THEORETICAL FOUNDATIONS

OF THE HISTORICAL SCIENCES

304 In the preceding chapters the role of history and historicality has pushed itself increasingly into the foreground. We shall now turn specifically to the question of theory in the historical sciences, applying the concepts that have been worked out up to this point.

Even today there is a widely held opinion that the historical sciences are concerned with the particular and individual, e.g. a certain personality, a particular state, a single epoch of art, etc., while the natural sciences deal with the universal, laws which are valid everywhere and phenomena which are always the same. Correspondingly, the methods of the former supposedly differ from those of the latter: It is said that the historian "understands", that is to say, he can make an empathetic contact with the details of human relations familiar to him, whereas the scientist "explains", and therefore refers the phenomena to general laws. This view, or one similar to it, has been adopted in particular by German philosophers and historians, such as Herder, von Humboldt, Dithey, Ranke, Droysen, Windelband, and many others.

In the past the views of these thinkers were often opposed by Anglo-Saxon authors. Recently this opposition has been reasserted. For ex-
305 ample, Hempel, Oppenheim, Gardiner, White, and Danto,[1] to mention just a few, have alleged that there are also explanations in the historical sciences and that general laws are used there as well. In this respect, then, all empirical sciences are supposedly the same.

On the one hand we have the philosophers of understanding, on the other, the philosophers of explaining. I shall begin by discussing their different standpoints, starting with the philosophers of understanding.

13.1 The Philosophers of Understanding

To begin with, let me present the view of these philosophers more fully. The position that the historical sciences are *only* concerned with the particular and individual is one which has often been superficially ascribed to these philosophers. This is, however, incorrect because the particular or special (*das Besondere*), which they lay such stress upon, is itself in a certain respect something universal. But this universality differs from that of natural laws in that it can in principle be changed by men—which is to say that it can be violated and therefore is only a historically limited universality. The fact that natural laws are themselves in part only human constructions, in the sense which we have already described, has no significance in the present context. Here we are not concerned with the conditions of the knowledge of natural laws, but rather with the notion that natural laws, regardless of how they may have come about, are *held to be* an immutable representation of nature, whereas a comparable immunity cannot be ascribed to the kind of universal with which the philosophers of understanding deal. And even if nature and its laws were similarly viewed as subject to historical change, people would still never be the agents of these changes when viewed from the standpoint of the object construed rather than from that of the construing subject. For example, a historical phase of physics is reflected in Newton's law of gravity; but whereas this must be viewed as something which no person could oppose, the same thing certainly does not hold for a law taken from the codifications of civil law. In the following we will focus only on this difference within the concept of the universal. 306

Having thus clarified matters, we can indeed stipulate that a particular state, a particular constitution, an economic system, a religious doctrine, or a style of art, etc., is individual and as such historically limited, as the philosophers of understanding have stressed; but on the other hand such things have something of the universal about them as well, that is, insofar as different individual phenomena of political, economic, religious life, etc., can be ordered within more comprehensive contexts. If I am not mistaken, there is hardly anyone among the philosophers of understanding who has denied such general forms of order and who has thus devoted himself to a radical nominalism. If they have in fact accentuated the individual in the historical sciences and have done this quite strongly, it is only because of a desire to draw attention to the historical uniqueness of these forms and to distinguish them clearly from the universality of the laws of the natural sciences.

However—and here my criticism begins—*precisely* what is to be understood by the universal is something the philosophers of understanding have not made clear. In this regard their ideas are more or less vague; and we certainly find nothing like an exact definition. Some of them speak rather hazily about organic and plantlike "wholes" (*"Ganzheiten"*), which are thought to comprise many things; others see the universal in terms of relational contexts of meaning or efficacy (*Bedeutungs- oder Wirkungs- zusammenhänge*) of life, etc.[2] In order to describe such obscurities, to circumscribe and to penetrate them, they must also conjure up special faculties of empathy, understanding, or sensing—even of divining.[3] 307

13.2 The Philosophers of Explaining
The philosophers of explaining are opposed to this view. Their opinion can be made clear by a very simple example. Suppose someone has lit a stove. This could be narrated in the following way: "Someone was freez-

ing, but he had a stove. And because people who are cold try to obtain heat, he consequently lit his stove." Obviously in this story[a] a sentence about an individual event (someone lit the stove) is deduced from premises which include a general law (all freezing human beings try to get heat). Now, according to the philosophers of explaining, every scientific explanation consists of such a deduction. It always hinges on a conclusion drawn from premises which, as the example shows, contain general laws. And they are convinced that such explaining is the important point in the histories written by the historians and that this explaining does not differ essentially from that of the natural sciences.

I agree—and I will show later why—but I am afraid that the philosophers of explaining, preoccupied with this insight as they are, have almost completely overlooked that kind of universality which is of special interest to the historian and which is essential to his work. Thus the philosophers of explaining concern themselves almost exclusively with general laws in the manner shown by our example. Doubtless, these also occur in historical explanations; but in fact the kinds of laws which these philosophers have in mind are laws of psychology, biology, and other sciences. On the other hand, the philosophers of understanding, as we will show, have indeed correctly seen that something else, that is to say something really historical, is principally at issue here; but their vision has been obscured to too great an extent by questionable metaphysical presuppositions.

13.3 The Specific Kind of Universal Belonging to the Historical Sciences

We are concerned primarily with an elucidation of that kind of universal which is specific to the historical sciences. Let me again begin with an example: Suppose a statesman has refused to order the murder of an opponent, even though this would have proved politically advantageous for him. This could be explained in the following manner: The statesman in question was an advocate of certain political principles. On the basis of these principles he believed that he had to attain a certain goal. He thought that eliminating his opponent at a favorable moment was the best way to achieve his goal. At the same time, however, he was an advocate of certain moral principles which he considered more important than political principles. Thus, because he thought that the murder of his opponent contradicted his moral principles, he refused to order such a crime.

Seemingly no one law of the kind "freezing human beings search for heat" can be found here; rather each sentence belonging to the premises refers to an individual event like "He was an advocate of . . . ," "He thought, believed . . . ," etc. However, viewed from the scientific perspective, this is deceptive since the law which makes the inference of this

a. The reader should bear in mind that in German the word *Geschichte* means "story," "narrative tale," etc., as well as "history."

explanation logically possible in any sense at all has been left out. This law consists in the assertion that human beings who believe, think, wish in the described manner, and who are in a certain situation like that of the statesman, will also act like him. Nevertheless, no one, with the exception of a very strict logician, will have noticed the omission of this law in the present explanation. In its abbreviated form, this explanation is perfectly understandable. The reason for this is that this law is of no interest at all to the historian here, because he focuses his entire attention on something quite different. I must, however, admit that this will not always hold true. It may well be that someone believing in a special *rule* (for example, in principles, as did the statesman mentioned above), according to which he has to act on certain occasions, will nevertheless not act in this way, because (e.g.) he is handicapped by psychological, biological, or physiological factors, etc. In such a case the historian will refer *expressis verbis* to a general law in the sense of the philosophers of explaining. 310

Very often, however, the historian will differ markedly from the scientist in his way of explaining things, as the following comparison may show:

POSSIBLE FORMS OF EXPLANATION

Historical Sciences	Natural Sciences
1. Someone was in a certain situation.	1. Something was in a certain situation.
2. At this time he believed in the validity of a certain rule, governing the way one must act in such a situation.	2. Whenever something is in such a situation, it will change according to certain laws.
3. Someone who satisfies premises no. 1 and no. 2 will act/will not act according to the specified rule because of psychological, biological, physical, etc., laws.	3. Consequently it has changed in accordance with the laws.
4. Consequently he acted/did not act in accordance with the rule.	

Here we see that what is properly essential to the historical explanation is no. 2 in the left column. No. 3, the law, will be omitted in most cases, even if it is not logically correct to do so. On the other hand, the scientists can never omit the law under no. 2 in the right column, since this is the point of greatest interest to them. 311

Even though I will discuss this in greater detail when dealing with axioms of the historical sciences, here I would like to explain further what I mean by general rules. These rules are nothing else than those already mentioned in chapter 8. The first things we encountered in the present context were ethical and political principles. Included in this category are, for instance, things like the Ten Commandments given in the Bible, the categorical imperative, and political guidelines as the general determinants of the political will (the Charter of the United Nations, the nationalization of industries, etc.). However, general rules also determine economic and social structures, even if these rules are not always written down somewhere and codified. The same is true of legal principles and of laws deducible from them. Furthermore, we find general rules in the arts and in the religious realm: For example, here we find rules functioning as laws governing the theory of harmonics and determining tonal systems, elements of style, forms of ritualized activities, etc. As already noted (cf. chapter 8), one could give as many examples here as there are various areas of life. At all times our lives are determined by rules which very often have the same rigor and exactness as natural laws. One can think of the rules we follow in our everyday behavior toward other people, the rules of etiquette, hospitality and good manners, the rules of driving, the rules governing business, money, and trade, the rules of behavior at work, and most of all the rules of language. Even when we are just playing a game, we submit ourselves to exact rules (the rules of the game).

Now and then the historian will encounter the ideal case described in chapter 8 in which such rules not only are codified but have even been brought into a strict logical and systematic order. In such cases, if the historian is a historian of science, his object might be a physical theory like Newton's; or if he is an historian of law, the object might be a code of laws. But one often comes across rules that have not been codified. In such a case the historian will first of all attempt to reconstruct the rules. Examples of such rules are things like the regulations governing barter in the ancient world, the basic principles upon which ancient Sparta was founded, or the lost plan for a battle that was obviously fought according to such a plan. In these cases we will rarely have something which suffices for a formal ideal of exactness, as I have already indicated; but in most instances of this type we find as much exactitude as is necessary to apply the rules in question within a particular set of circumstances.

The results of what I have said thus far can be summarized as follows:

First: The universals which are central to the historical sciences are rules. In my opinion it is pure mystification to see in this, as do the philosophers of understanding, vague organic *Ganzheiten, Bedeutungszu-sammenhänge,* or something of this kind.

Second: These rules belong to the past and have a historically limited efficacy. This being the case, I must again oppose the philosophers of explaining who are unable to adequately divert their attention from laws with a historically unlimited efficacy and thus fail, in my opinion, to consider what is intrinsically historical. Of course, even the historian uses general laws, as I have already noted; but to the extent that he does this, he is acting as a psychologist, a biologist, or a physicist, etc., whereas he is a historian only by virtue of the fact that he refers to the kind of universals of which I have just spoken.

We will now look at two points which bring out the degree to which the philosophers of explaining have erred and which will serve to enhance the present critique of these philosophers.

As far as I can see, explanations which lack the kinds of laws described above are considered by these philosophers to be merely "explanation sketches" or "quasi explanations." But it seems to me that such phrases are misleading because they give the impression that the historical sciences are in some way flawed, that they are particularly vague, and that this is what principally distinguishes them from the natural sciences. For example, when someone says that he has just taken a pill because he has been suffering from a headache, I think that in most cases no one would seriously call this an "explanation sketch." Whether or not such a phrase is appropriate depends upon particular conditions. Just as in such everyday explanations, most historical explanations do not require further evidence, and are clearly and unambiguously understandable. Too strict a demand for completeness might indeed be more harmful here; it might unnecessarily complicate matters and in the end create real unclarities. And this holds for the natural sciences as well. Moreover, those philosophers who have fixed their attention solely on the laws present in historical explanations have also been led astray by considering rules to be laws, since they are blind to rules, locked into their orientation as they are. For instance, they speak of economic laws, even though closer examination reveals that these are institutional norms, like the rules of free-market economy, the rules governing the gold standard, etc. Or we might take W. L. Langer's attempt to interpret certain medieval occurrences with the aid of laws of psychoanalysis, an attempt that has been taken up by some philosophers of explaining. Langer uses psychoanalysis to trace the origin of several motifs in late medieval art back to a general trauma evoked throughout Europe by the devastation of the plague—motifs like the dance of death, depictions of hell, the Last Judgment.[4] However, he completely overlooks the point that such an occurrence could have had the observed effect only because the people lived at the time in the spiritual-cultural world of late medieval Christianity and its art. Depictions of hell and of the Last Judgment could never have been evoked by the plague that raged through Athens during the Peloponnesian War. But the foundations and basic structures of late medieval Christianity and its art are not psychoanalytic laws; indeed, they are not laws at all, but rather rules of a historical era.

Thus I generally regard the cultural, political, social, and religious worlds in which the historical persons have lived as much more important for the most part than the psychological laws and so-called dispositional properties so often applied today in the philosophy of history. In contrast to this, the philosophers of understanding have, I believe, quite correctly seen that in the historical sciences a different kind of universal from that of the natural sciences is the focal point. On the other hand, however, they have not realized that this kind of universal, when viewed only with respect to its logical form, does not differ from natural laws, since like them it consists of rules.

13.4 The Internal Connection
between Explaining, Understanding, and Narrating

Explaining is carried out under the rules discussed above. The central point in this, as the philosophers of explaining have said, has to do with a mode of inference. Understanding, regardless of how it is defined, may accompany and aid explaining, but it is not necessary for explaining. In any case, the historian explains; whether or not he understands while explaining is another question. Thus it is often the case that some aspect of the behavior of past cultures is explained, even though we have no direct access to these cultures. But here one can question whether understanding means anything but explaining with the help of a context of rules or laws that is merely especially familiar to us, that contains either an element of our own reality or a reality that has become "personal" to the historian through his constant contact, study, etc. (as with the historian who is so immersed in the past that he can feel and think like an ancient or a medieval person, etc.). The basis for what is foreign, or even incomprehensible, in cultures and peoples that are distantly removed from our own world surely lies in our incomplete knowledge of their behavior or in the difficulties encountered in trying to order their behavior within the spectrum of rules which are familiar to us. Here it should be noted that understanding is not to be equated with condoning or sympathizing. Given sufficient familiarity with the context, even a crime can become understandable; but it certainly need not correspondingly be condonable.

Viewed in this way, the assertion that nature—as what is foreign—can merely be explained but not understood becomes meaningless. A large number of natural occurrences are in truth as familiar to us as human life; and our knowledge of nature, within which we act quite effortlessly and spontaneously, is no poorer than our knowledge of human life. We can recognize this readily in the cults, mythology, art, and poetry of peoples and cultures who have not disguised their view of the environment to the extent that we have. We first encounter what is foreign in nature when its indifference to human aims becomes evident, and especially when it becomes the object of a kind of observation which deliberately excludes our everyday interaction with it, as in the natural sciences. The impossibility of separating the natural from the human world in certain contexts gives, I believe, the clearest evidence that understanding cannot be referred exclusively to the human dimension, and that it basically rests only

316

upon a complete familiarity with a comprehensive context of rules or laws.

Now, some thinkers believe that the essential characteristic of the historical sciences is lost to those who make the concept of explaining such a central issue. These thinkers say that the historian is not so much engaged in explaining as he is in narrating. But I believe that any expla- nation in the historical sciences is also a narration and that in this field we can scarcely separate narration from explanation. The explanation of the statesman's actions in our earlier example can be used to illustrate this point, since there we can see that the explanation is quite obviously also a narration. Danto, in the book already cited, has pointed out quite well the tight interconnection which exists between explanation and narration (cf. Danto, *Analytical Philosophy of History,* chapter 11, "Historical Explanation: The Rôle of Narratives"). There he remarks that every narrative depicts a transition from the events present at its beginning to the events present at its end. Thus, according to Danto, narration can be said to take the following basic form:

1) x is F at time t_1,
2) H happens to x at time t_2,
3) x is G at time t_3.

Hence no. 2, the middle part of the narration, explains how the transition from no. 1 to no. 3 took place. The general law is indeed missing in this explanation, but it is indicated; it can be distilled, so to speak, from the explanation "An F to which H happens changes to G." This does not melt into a "miserable tautology" à la Hegel, as the schema of possible forms of explanation presented earlier in this chapter shows. For in the first place it cannot be said that the law introduced in the third premise of this schema is always trivial, as I have already mentioned (i.e. especially not in complicated psychological or biological situations); in the second place, even if the law is trivial for the historian, it is still not devoid of content; for there still exists a rather opaque psychological connection between volition, believing, and acting. However, here we cannot take the time to discuss this at greater length.[5] According to Danto, then, a strictly deductive explanation and a narration are merely two different forms of explanation, and the one can be translated into the other. Admittedly it must be noted that narratives often illustrate change taking place over a vast span of time, so that the middle part consists primarily of mere single steps in the above given form (which Danto therefore calls atomic narrative). In conclusion Danto stipulates the following essential characteristics for a coherent narrative (and this is the type of narration which is to be expected from an historian): (1) It concerns a transition in which something serves as the continuous subject of the transition. (2) It explains the transition of this subject. (3) It contains only as much information as is needed for (2). Here again, the analogy to deductive explanation becomes clearly apparent.

13.5 The Concept of "Theory" in the Historical Sciences

Following this effort to further clarify the specific type of universal belonging to the historical sciences, let us return to the concept of historical theory that has already been discussed in chapters 8 and 11. Everyone knows that there are theories in the natural sciences. We have various theories of light, gravitation, microphysics, etc. Strangely enough, however, this concept is seldom used in the historical sciences, or at most only very occasionally; and in no case, as far as I can see, is it used systematically and with full consciousness of what might be meant by it.

The aim of theories in the natural sciences is, among other things, to explain and classify a special group of natural events within a context of natural laws that is as comprehensive as possible, and to reduce the events to this context. We can speak analogously of theories in the historical sciences. For "natural laws" we can substitute "rules for a certain realm or sphere" (e.g. the Roman judicial system), rules which are selected in such a way that we can deduce as many other rules from them as is possible and appropriate for this given sphere. Thus we can say that these theories also serve to explain and classify a special group of events, which are now historical events, within a context of rules that is as comprehensive as possible, and to reduce the historical events to this context.

I see a close connection here to Max Weber's "ideal type," even though Weber was apparently not aware that this "type" had to have the form of a theory. This can be clearly seen in a rather detailed quotation from his essay "The 'Objectivity' of Sociological and Socio-Political Knowledge," which can also serve as an example of historical theory. First, Weber says that we make a thought-picture of the occurrences of the market in bartering forms of society, in free competition, etc. Then he goes on to say: "This thought-picture welds together certain relations and occurrences of historical life into a . . . cosmos of conceived contexts. . . . The relation between these contexts and the empirically given facts of life consists simply in our ability to make the uniqueness of a context . . . understandable in terms of an ideal type; for instance, wherever . . . we can establish and assume in reality . . . occurrences which are dependent upon the market."[6]

According to Weber, then, this is how an idea like that of urban economy in the Middle Ages is constructed, and thus how an "ideal type" comes about under which single phenomena might be combined into a unified thought-picture. I believe that his understanding of this was basically correct, even though he neglected the fact that this synthesis or combination is nothing other than a theory of rules, for the idea of urban economy in the Middle Ages can only consist in such a theory.[7]

This example also serves to clarify the question of what it is that a historical theory describes. It describes a system in history, just as a theory in one of the natural sciences describes a system in nature. This means that it imputes the past efficacy of a system of rules to a group of historical phenomena, whereas a theory in the natural sciences presupposes the efficacy of a system of laws in a group of natural phenomena. Once again it is evident, just as we can conclude from chapters 8 and 11, that a scientific, that is, a theoretical, consideration of history must be related to historical systems. Thus, for example, a theory of the medieval 321 market regards the economy of this time as determined by a system of rules which is described by this theory, whereas a theory of optics views the phenomena of light as determined by an immutable system of natural laws which it describes.

Now, some may raise an objection to this use of the concepts "theory" and "system" in the historical sciences, namely, that along with such concepts a rationality and logic is imputed to history which it does not possess. This amounts to saying that history cannot be molded into systems, that most of what happens in history is too indeterminate and occurrences are largely governed by passions, errors, delusion, and contradictions. For instance, Schopenhauer writes: "The subject-matter of history . . . is composed of the ephemeral interactions of a human world as mobile as clouds in the wind, which is often completely transformed through the most minor coincidence."[8] "What history relates is in fact only the long, difficult, and confused dream of humanity."[9] If this were in fact true and true to such an extent, then it would be impossible to write history; indeed, history would not even exist. Concerning this, I have already pointed out that historical systems often do lack logical consistency or are insufficiently clear; and furthermore this also holds for the interpretations of these systems and the conclusions drawn from them by individuals acting within history. But if the systems are faulty, then this must be reflected in their theory; and if the interpretations are faulty, then this must from time to time be explained by utilizing means that are not inherently historical, but rather, for example, psychological. For as I 322 have already said, both historical systems *and* systems of nature are active in history. It is because of this that people often grasp at idealizations, as did Max Weber when he used the term "ideal type"; and often the attempt to bring things into some sort of order fails. But all this already presupposes not only that such attempts cannot be ignored by the historians, but further that they are an invaluable heuristic tool for the historian's science or, as Kant would say, that they serve as a regulative idea. Anyone who begins by refusing to make such attempts is also refusing to write history scientifically. Citing Kant once again, this would be "lazy" reasoning.

In addition, however, I would again like to point out the danger in underestimating the logic of historical processes. I have already indicated that our lives are regulated by a wealth of different rules, and this pertains even to everyday details. When these rules are disrupted, they are for the most part merely replaced by new ones. We know that even madness has its methods.

13.6 On the Justification of Theoretical Principles in Theories of the Historical Sciences

Thus I assert and maintain the following: We have theories not only in the natural sciences but in the historical sciences as well; and theories in both disciplines have the same logical structure. However, because of this, and contrary to widespread opinion, we face the same epistemological problems in both kinds of theories insofar as these problems are based on the logical structure of theories in both cases.

Each theory of history is also necessarily based on certain principles, and therefore the question of the justification of these principles must again be answered. In the first place, we have epistemological principles of the most universal kind, as for example the principle of retrodiction. We make use of this in all empirical sciences, as well as in daily life, whenever we draw conclusions about past developments from present ones. However, principles belonging specifically to the natural sciences are also important for the historical sciences, since the latter have to make use of physics, astronomy, biology, etc., as auxiliary sciences. This occurs, for instance, whenever the age of an archaeological find has to be determined, or when the authenticity of a document has to be investigated by using genealogies, etc. Finally there are principles belonging specifically to the historical sciences. These, however, we can submit to the general categories developed in chapter 4, that is, to the same categories which pertain to the natural sciences. In both fields we find principles which we could call axiomatic, judicative, and normative. I now want to elucidate these principles more precisely and at the same time to point out the epistemological problems inherent in them. These problems do not depend on the specific content of the theories; thus it makes no difference whether this content belongs to the natural or the historical sciences.

13.7 Axiomatic A Priori Principles in Theories of the Historical Sciences

Axiomatic principles are understood as those which make up the core of a theory. In the natural sciences these are assumptions about the fundamental laws of a natural system (for example, the Schrödinger equations). In the historical sciences, however, these are assumptions about the fundamental basic rules of a historical system. Only the latter will concern us here. What constitutes these principles was indicated earlier in the context of a more general discussion of what constitutes rules for the historian concerned with explaining. Let us now examine this more closely by introducing several examples taken from the history of historical writings concerned with ancient Rome.

These historical writings can be viewed as a history of theories about the Roman state and its culture. This means that the authors of these writings set up fundamental structures by means of which individual events can be explained and diverse phenomena assembled under unified headings—rules and concepts as rules. Thus Gibbon in his *History of the Decline and Fall of the Roman Empire*[10] interpreted the historical drama in the light of general cultural (*geistigen*) structures dominating Christianity and the end of Classical Antiquity. We can see this even better with Niebuhr, whose starting point for his *History of Rome*[11] was the social foundation of the Roman Empire and its agricultural constitution. He too tried to bring an abundance of historical material under the order of a general systematology and to gain a comprehension of this material through the use of systematic principles. Mommsen did the same, though he far surpasses Niebuhr in knowledge of legal matters and is thus better able to found his history in a thoroughgoing manner upon systematic principles[12] New aspects can be found in Rostovtzeff,[13] who develops a history of Roman economy and society by using a few basic concepts. More recently Heuss has attempted a rough identification of Roman domestic policy with the history of the Roman Constitution, depicting the former as an embodiment of the latter.

Heuss writes: "Instead of illustrating a great variety of events, a transparent classification of facts should be given, which might facilitate understanding and provide us with a guideline for orientation. . . . The objective principle for this was the concept of revolution. This made it necessary to classify the material according to the phases of the revolutionary process and to make its different structures as clear as possible."[14]

However, we also find theories present in the consideration of individual events in Roman history. For example, Roman expansion is explained on the one hand as based on a Machiavellian principle of the pure will to power, but on the other, and this is especially true with Mommsen, it is seen to rest on the principle of securing stability in an ever-widening area. In Roman foreign policy we also find the practices of legally defending declarations of war and the strict ritualization of such procedures interpreted as things which are derivable from Rome's conservative constitution, since these practices were always intended to identify the enemy as a threat to the existing and traditional belief in justice (*Recht*). Further, the various theories concerning the basic principles of the *Optimates* and the *Populares* are of fundamental significance. Thus one group sees a class conflict in this opposition, while another group understands this to be merely a constitutional conflict (i.e. whether Rome should be governed by the Senate alone or the Senate in cooperation with popular interests represented by the *Comitia*). Finally one should mention Meyer's attempt to derive all civil, imperial, and foreign policy, in general the entire scope of Augustus's political activity, from one constitutional principle.[15]

Now, such axiomatic principles in theories of the historical sciences are a priori principles like those belonging to theories of the natural sciences, even though such a statement might seem strange at first glance. They are a priori principles in the sense that they make possible any knowledge of facts whatsoever, while, on the other hand, they can never be verified or falsified directly by facts.

In order to demonstrate this, I shall begin by considering the view that a historian can only understand a document by knowing the legal, economic, or social conditions which prevailed at the time when the document was produced. But the question is then: How is the historian to learn about this? The only answer is that he must again use documents and other sources. With the help of these he will try to order the variety of phenomena in terms of certain connections and deduce them from certain principles—that is, he will construct a historical theory which fits his
327   sources, so that he will be able to interpret the document in question, first, so as to arrive at any facts at all, and then to explain these facts with the aid of the documents thus interpreted. Hence the historian does not proceed any differently in his work from the natural scientist. In both fields of science, the individual facts will only be given in the light of a theory. Or, as we have said, facts depend on theories. Therefore a theory is indeed "the condition of the possibility of experience."

On the other hand, this theory will be tested by experience. For example, the conception one develops concerning the principles of Roman law in order to explain certain historical events or interpret certain documents will be either proved or disproved by, among other things, documents. The rules developed and projected as the governing factors for the geometrical style of Ancient Greece can be corrected if we bring into play the extant pottery of that time. We can find out if the activities of Napoleon correspond to the political objectives imputed to him. With regard to all this, it is normally said that an interpretation either does or does not succeed. Here I have merely translated this means of expression into the terms introduced in the course of the present discussion. But even though experience does come into play here in the above-mentioned manner, no theory will ever be verified or falsified empirically in an unambiguous and absolute way. For, as I have already shown, the sources and facts used in testing theories themselves presuppose and depend upon theories of historical science. For this reason, each verification and each falsification has to be something hypothetical. Furthermore, the logical scheme of a confirmation consists in the harmony between statements

deduced under certain conditions from supposed theories and the interpreted facts. However, for logical reasons, as we have already seen on
numerous occasions in the previous chapters, the confirmation of the $\quad$ 328
deduced statements does not say anything about the confirmation of their
premises—in this case the axiomatic principles. Thus it is impossible to
give a direct empirical justification for them; on the contrary, we have to
construct them a priori.

13.8 Judicative Principles

Let us now turn to the judicative principles. By these I again mean those
principles which make us reject or approve theories on the basis of interpreted facts. Now, because this cannot occur in a strictly empirical
way in either the historical or the natural sciences, as I have just demonstrated, there must then be certain rules governing the manner in which
this is to occur. For instance, we will have to decide whether the factual
assertions contradicting a theory are to be accepted or, even more precisely, whether the theoretical presuppositions of these factual assertions
are themselves to be accepted. If they are accepted, we will have to decide
next whether the theory is to be regarded as falsified or whether these
findings should simply be attributed to some special, atypical circumstances. In addition, we might hold that a theory must be rejected if it
can only be defended by ad hoc hypotheses; or, vice versa, we might
accept such hypotheses under certain conditions and circumstances. We
might be convinced—for whatever reasons—to adhere to certain axioms
no matter what the case, or to acknowledge a theory only if it has a wider
and more comprehensive content than other theories; if it explains something which is as of yet unexplained; if it aids in revealing something
previously unknown; etc. Basically, all this has already been said in the $\quad$ 329
previous chapters, but there it obviously all referred to the natural sciences. However, none of these rules of falsification and verification, regardless of what we might think of each one individually, can be based
on experience, because experience presupposes them. For example, these
rules determine whether a fact can be considered a test of a theory and
whether, in case of harmony or disharmony between the fact and the
theory, this fact can be considered a confirmation or refutation of the
theory. After we have made this decision in favor of some set of a priori
rules of this kind, historical (or natural) reality will then tell us whether,
under these conditions, it says yes or no to these theoretical constructions
we have built up in this way. (I will deal with this more extensively later
on.)

13.9 Normative Principles

We come then finally to the normative principles. Their very name already
expresses their a priori status. They tell us what it is that can belong to

a scientific theory in any sense at all. With regard to a theory of the historical sciences—and only these are now under consideration—we will expect to make use of historical geography, chronology, genealogy, paleography, heraldry, sphragistics, numismatics, and other disciplines. We will expect a historical theory to use a large number of such aids, which also fulfill certain normative requirements and are therefore called auxiliary sciences. Within them we even find a number of natural sciences like geography, astronomy, and biology. Of course, the greatest importance is placed on the use of sources; and respectable scientific methods of critical selection and judgment have been developed especially for this purpose. Finally we might add that reference to any kind of metaphysical power or force, as, for example, to a divine providence, is not acceptable.

13.10 The Relation between the A Priori and the A Posteriori

Let me now try to explain the relation between the a priori and the a posteriori in a somewhat more detailed manner with the aid of a very simple model.

1. a is F (Fa) (T_1)
2. Whenever F, then G (T) $\quad T_3, T_4$ $\qquad S_1$

3. a is G (Ga) (T_2)

This schema expresses an inference in which Fa and Ga are singular statements, and premise 2 is an axiom of the theory T. T_1 and T_2 represent theories with the help of which Fa and Ga are respectively given. T_3 stands for the normative theories according to which T, T_1, and T_2 are accepted as regards their form. T_4 represents the judicative theories according to which Ga is viewed/not viewed as a confirmation of T. Finally S_1 is the set of all the theories T to T_4. Where do we then find the empirical element within this model?

The empirical element surfaces in the following manner: If we presuppose S_1, then we get a particular result R_1, namely, Fa, Ga and thus the confirmation of T. This is empirical because nobody could know it a priori. If on the other hand we now substitute S_2 for S_1, then we might possibly obtain a different result R_2. This too is empirical, because it might just as well turn out the other way around; that is, S_1 might yield result R_2 and S_2 might yield result R_1. Thus no single part of this model is empirical as such, neither the different S-sets nor the singular statements; the only empirical factors are the different hypothetical metastatements of the kind "If we presuppose S_1, then the result is R_1," and "If we presuppose S_2, then the result is R_2," (cf. chapter 3).

Now, if we choose to say that theories can be confirmed or refuted empirically, we are merely speaking in an elliptical manner. Actually, these theories, along with their principles, are constructed in an a priori manner, since on the one hand they first make experiences possible while on the other they are not directly testable by experience. Such a test can only be carried out under conditions which are set up by an S-set. For this reason the unpredictable result of such a test also depends on the S-set.

As I have already pointed out, the singular statements in the model are not, strictly speaking, empirical, since they also, like the principles, are not given directly by experience. However, this can only be considered true insofar as any statements about interpreted objects of experience have a theoretical content which exceeds the empirical content. Moreover, this theoretical content is itself a part of the S-set. Therefore in the model a singular statement expresses experiences under certain conditions. The S-set, on the other hand, expresses only the conditions; and it is these alone which I call a priori.

Consequently, the a priori element cannot be eliminated, and this holds for the historical sciences as well. As such, however, this a priori element always stands in need of a special justification. But before going into this *quaestio juris,* I should like to supplement my treatment of this topic with some important points.

13.11 The So-called Hermeneutic Circle
The model just given shows that the so-called hermeneutic circle, which   332
is discussed so often today, does not really exist. First of all, this erro-
neously named "circle" comes up not only in the historical and human sciences, but in every empirical science, since the relation between a priori assumptions and facts interpreted with their help is principally the same everywhere. Hence we are not dealing with something inherent to the historical and human sciences alone.[16] Now let us look once again at the schema presented by our model, bearing in mind that this schema applies to all possible empirical sciences. To get a clearer idea of what is involved in this so-called circle, let us assume that $T = T_1 = T_2$ (so that "When-ever F, then G" is merely one of several axioms of T). Let us then further assume that we in fact obtain Fa and Ga by presupposing T, and finally that conversely T is confirmed by Fa and Ga. But in such a case there is then no circle, since, as I have already demonstrated, only experience will show whether we arrive at such a result, that is, whether Fa, Ga and the confirmation sought will result under the given presuppositions of an S-set. We could never obtain this result from the theory alone. Conse-quently, it is only conditionally true to say that we only get out of things what we have put into them; experience enters in, even if it is already   333
"preformed" in a sense by a theory, as shown with S in the model given above.[17]

If Ga is a fact which is as yet still unknown to the historian, then he will be able to predict it on the basis of premise 2, just as the scientist usually predicts in an experiment. Correspondingly, the historian will then also find supporting or refuting evidence in subsequent findings, discoveries in archives, excavations, etc. But if Ga is a fact which is already known to the historian, then he could interpret it in a way with the help of his theory, or relate it to other facts, for instance, Fa, which would then be seen as the confirmation or falsification of the particular theory held by the historian.

13.12 The Explanation of Explications and Mutations of Historical Systems and the Explanation of Meanings

Up to now we have only been dealing with the historian's explanation of events and facts within space and time. Such events and facts were illustrated by the example of the statesman who acted in a way in accordance with rules. But the historian must also explain the genesis of rules themselves, that is, the coming about of ideas, opinions, images, practices, styles, etc. Constitutional and legal history, economic history, art history, etc., provide countless examples of such explanations. All so-called cultural histories or histories of ideas are like this. However, how does this occur? As demonstrated in chapter 8, there are only two ways in which historical systems, because of their form, can be subjected to movement or change. This can be done on the one hand through explication, and on the other through mutation. By the "explication" of a system, I mean its inner development within the framework of its basic rules, whereas "mutation" signifies a change in these basic rules and thus the genesis of a new system. The question posed above can thus be more precisely stated in the following form: How does the historian explain explications and mutations?

Let us begin with explication. Here we are always dealing with rules which develop out of other rules. As already stated, the ideal case is that of a theory in physics; an example is Newton's theory, from which more and more theorems are derived and for which more and more applications are discovered. Such a theory together with its explications can indeed become the object of a historical investigation, as in the history of science. But certain developments in the histories of constitutions, laws, politics, economics, art, etc., also admit of derivations of this kind. These derivations come about whenever conclusions have to be drawn under particular conditions within a given context. The judge, the businessman, the politician, and the scientist all act almost daily in this way. The game of chess is analogous: The basic rules are set; but different games, different opening moves, strategies, etc., all unfold in accordance with the situation. Each move can be derived from the basic rules; but whether the move will actually be made, whether it will appear in a book on chess as a particular strategy in a particular configuration, can only be explained through the enactment of the game of chess. For here we must not confuse the logical possibility of the derivation, the fact that it is implied in the basic rules, with its actual occurrence.

Naturally, the schema for the explanation of explications of historical systems does not differ fundamentally from that already given for singular or individual facts, since scientific explaining always has the same form. Thus we can outline this schema as follows:

1. A rule-set R is given for some person (or group of people).
2. This person finds himself in a particular situation for which he has to set up a rule that can be obtained or derived on the basis of R.
3. He believes that this applies in the case of R'.
4. Anyone who fulfills premises 1–3 will/will not set up R' because of psychological, biological, physical, etc., laws.
5. Consequently he does/does not set up R'.

As an example, we can use the apearance of particular motifs in late medieval art which was discussed earlier. The people of that time lived within the rule-set of this kind of art. They sought to develop new art forms from the ones already given, forms which were appropriate to their traumatic situation, dominated as it was by the catastrophic plague raging through Europe. They believed that they had found these forms in the motifs mentioned earlier. Hence these became a new rule that could be understood in terms of the old rule-set. Incidentally, with this example we can see clearly that the explicating derivation need in no way be logical in the strict sense of the word. We are dealing here with rules that are artistic forms and figures. We can consider their systems to be in part analogous to the calculus of a game. The clarification of the various kinds of explicating derivations is a broad field for future investigation in many areas. In my opinion the present system-thcoretical approach is a beginning point and can serve as a helpful guide for such investigations.

Let us now turn to the consideration of the mutation of a system. Here we are concerned with a change in foundations. Hence we do not remain within the system as with explication. Rather we take up a critical distance from it; we remove ourselves from it; we speak about it and treat it as an "object language"; we place the presuppositions and basic premises of the system at our disposal, and then we change them. Obviously our only foundation for effecting such change is to be found in the fact that we proceed on the basis of different presuppositions; we consider the old system critically from the standpoint of a different system, and we try to assimilate the two. In its formal sense, this process consists in the derivation of one system from another, whereas in explication, as previously discussed, we merely derive a rule from other rules within the same system. Viewed scientifically, nearly all the radical reorganizations in history have taken this logical form. Theoretical systems alter practical systems and vice versa; we find a mutual determination existing between

336

337 political, economic, scientific, social, artistic, and religious systems, etc. Such relations are produced everywhere; one system stimulates, one influences and alters the other. Thus the historian explains a mutation by construing this process via "reconstruction"; on the other hand, the logical schema for this is analogous to the schema of explication. Hence we need not discuss this schema in any further detail.

What I have just said about the explication and mutation of historical systems I call the *logical sense* of what has been commonly called in a very misleading way the "history of ideas" or "cultural history." Such history has long been described in literary, artistic, religious, economic, and legal history—in short, in all cultural history. It is misleading because allusions have been made to the development of ideas, or even spirit (*Geist*), where what was in fact the case was a simple development and setting up of rules. We have no need of such ballast from idealistic philosophy (ideas, spirit, etc.) when we are dealing with the rules according to which people act. To give an example: It is difficult to bring the rules of industry (which will be discussed in chapter 14) or of a football game into line with such complicated philosophical assertions, even though both are quite significant phenomena of modern culture.

Let me once again refer to the limitation of historical explaining with the aid of systems. First of all, we could pose the idea of a continuous but thoroughly entangled "family tree" of historical systems. This family tree could be described as an intertwined mesh of linked derivations. However, we would have to take into account, as I have said earlier, that
338 nonsense, contradiction, and even madness can be highly effective forces in history, forces which destroy any logical continuity. Furthermore, fundamental problems bar the actualization of this idea of logical continuity, problems having to do with the way in which the systems are viewed in terms of the theory of knowledge. For regardless of where these systems are applied, they can never be so compellingly grounded by experience or reason (as we can conclude from the relation described earlier between the a priori and the a posteriori) that we could eliminate a certain measure of spontaneous creative activity at their base, a creativity to which no kind of compelling insight can give a "final ground," as so many philosophers say or would have us believe. Once we accept this, it follows directly that historical systems can occur in the form of either continuous or discontinuous sequential chains, depending on their own characteristic mode of spontaneity. One can break off such chains and provide a new beginning point, since each of them is construed as something "floating" which never has a basis in *absolute* experiences or insights of reason. On the other hand, from our discussion about mutation it becomes apparent that this is only possible within the limits of a more comprehensive context, something which will become even more evident in what follows.

Up to now I have only been speaking about explanations of facts; in this I included the genesis of rules according to explication and mutation. *Explanations of meaning (Bedeutungserklärungen)* are, however, just as important for the historian, e.g. explanations of meaning which refer back and rest upon the sense (*Sinn*) of words. Hence these explanations also disclose the sense of the original sources to us. Explanations of meanings thus precede those of facts, because facts must first of all be established, that is, they must be taken from documents before they can be explained. I refer to this point again here, although it was discussed in section 13.7 above.

Let us consider another simple example. In a short story entitled *Averroes' Search,* J. L. Borges writes of Averroes that at first he did not understand the meaning of the words "tragedy" and "comedy" in Aristotle's *Poetics* and *Rhetorics*. No one in all of Islam knew what these terms meant. But suddenly Muezzin's prayer moved Averroes to write the following: "Aristú (Aristotle) calls tragedies panegyrics, and comedies he calls satires and anathemas. An abundance of superb tragedies and comedies are contained in the Koran and the sacred Mohallas."[18]

Averroes thus gains knowledge of a series of facts with this interpretation, for he uses it when reading Aristotle's history of the ancient theater; and with these facts he learns something about Aristotle's aesthetic rules of drama. (He constantly errs, but that is another matter.)

The analysis of explanations of meaning certainly does not seem to me to have been completely clarified; but two things can be inferred from this example which will suffice for our present analysis: First, there belong to explanations of meanings theories which consist in the setting up of definitions as general rules; second, we can test these theories empirically by finding out whether they agree with the textual passages in question. Only rarely will we actually be able to establish such agreement in a strictly formal manner; for this reason a broad spectrum of interpretation is always possible. Thus it seems that laws play no role in the explanations of meanings themselves, although they can certainly make an appearance here as well—for example, if one bases such an explanation on a special (e.g. psychological) intention of the author. But be that as it may, the historian cannot explain meanings without using a historical theory; thus even the explanation of meanings falls within the total family of questions being treated here.

13.13 The Justification of Theoretical
Principles within a Historical Situation

We can now finally turn to the basic problem already indicated to be common to all empirical sciences and thus to the science of history as well. This is the problem of the justification of the a priori principles upon which all these sciences are based. If there is then no absolute, no transcendental justification for them—something which I take for granted here—and, further, if no empirical justification is possible and if the condition described earlier as one of "floating" is not to be interpreted as pure arbitrariness, then we can only come to the conclusion that the foundations of these principles are taken first of all from other theories or, stated more generally, from other historical systems with which the historian is familiar. (This was pointed out earlier, in the discussion of system mutations.) The historian will try to connect all the various systems which make up his world, and in this way build up the most unified and the most comprehensive possible system of relations. He will try to

341 eradicate or at least to detect contradictions and any type of obscurity. This means that his chief intention is one which is directed toward bringing the historical material into this comprehensive connection and harmonizing its axiomatic, judicative, and normative a priori principles with the other principles present within this comprehensive totality. Therefore the historian deduces his principles from those of other fields or areas of life, assuming that these other principles are already justified by some reasons, and then he makes use of them in his own field of science. It may be helpful to give a few examples of this.

In the Enlightenment, when the science of history arose by separating itself from theology and dogma, the various influences a historian was exposed to were already well known. Thus the precursors of the German historical school, Schlözer and Rühs in Göttingen, pointed to the importance of other sciences for the historical sciences and also to their connection with political, social, and other such factors.[19] This they could do all the more since they quite consciously took what I call judicative and normative principles from the preexistent and extensively developed critical methods which had been established in classical philology and biblical research. In particular, they regarded the critical edition of the New Testament as the ideal. Schlözer speaks of comparing manuscripts, eliminating errors from a historical text, checking the text thoroughly in order to detect interpolations and forgery, uncovering the sources used by the authors, etc. Here we also find reference to the fact that the development

342 of such methods had its source in the widespread religious controversies which so strongly dominated the world in the period following the Reformation. Correspondingly, Gatterer, who was the first of the Göttingen school, connected the transfer of critical methods of scientific research to the field of historical science with different legal and constitutional questions of the highest political importance. Thus history becomes a science at a time when a normative concept of science had already been established in other fields. Indeed, on this occasion, this concept is even expressly transferred from the established fields to this new one.

However, it was not only biblical criticism and classical philology which aided in the development of this occurrence; the natural sciences along with a general theory of experience which was developed by them and free from any theology also played a decisive role here. This is already obvious in the work of Bayle (long before the Göttingen school arose), who applied this theory to the sources of historical facts, and thus developed a critical method for the historian.[20] At one of the high points in the development of the historical sciences, Webb could even speak about Ranke as someone who had changed the lecture hall into a laboratory using documents instead of retorts.[21] Most important, however, the historical sciences took over and incorporated axiomatic principles of the natural sciences. While this might seem a strange statement at first glance, it is shown quite pointedly by the introduction of the concept "law" into the field of history. Voltaire, who was one of the first to attempt to do this, expressly stated that it was his desire to write a historical work analogous to the physics of Newton.[22] The manner in which he interpreted the described events, the principles he considered as efficient causes in them, the systematic structure he believed these events possessed—all this was completely determined by the idea he had formulated in his mind. (The extent to which he was or was not successful in this attempt is of no importance in the present context, nor need we ask whether he understood Newton correctly.) Montesquieu may also have had similar ideas in mind, particularly since he deduced his "laws" in part from natural conditions, especially from climatic conditions.[23] However, in this regard we must admit that the direct and conscious connection with the natural sciences was in part lost later on. We even find that these first beginnings of enlightenment were rejected as the way *les philosophes* write history, something which had to be replaced by the *érudits*. But who can deny that one of the most famous of the *érudits*, Gibbon, does in fact transfer the axioms of rationalism, which arose together with the natural sciences, to his own historical work? This he did not only by looking at all events in terms of the "natural reason," thus explaining and interpreting everything from this perspective, but what is even more by trying to describe these events with the means used by the Enlightenment criticism of Christianity. Of course, today it has become rather commonplace to see the historical sciences not only in their connection with other sciences, but also, to use my terminology, in relation to a whole variety of systems in general within which they are embedded at any given time (for instance, in relation to the systems of politics, economics, social structure, technology, etc.). Such matters have acquired a particular importance today since we have once again started to take up the ideas once prevalent in the Göttingen school and propounded by Lord Acton, namely, to concentrate on the history of the science of writing history.[24] While this is indeed taking place today, nonetheless for the most part the problems bearing on the theory of knowledge and on the philosophy of science which pertain to this are generally almost completely ignored or over-

343

344

looked. In this respect, we must always bear in mind that it is not sufficient simply to detect such interrelations; rather we must also come to recognize the justifications for the a priori elements which lie hidden here and to see all the attendant questions raised by these justifications as well.

The world of theories and historical systems from which the historian gathers the justification of the principles for his theories, or by means of which he enters into a reciprocality of foundations, is what makes up his historical situation, in the sense given this term in chapter 8. There is no constantly valid, self-grounding or absolute reason and no pure, uninterpreted absolute experience which might serve to cut him off from his historical situation and loosen or dissolve his connection with it. The picture of history he develops is in itself something historical. He constantly makes use of a priori elements. He has to base his thoughts and experiences on these a priori elements as he moves forward, while at the same time never being able to make the a priori itself in its whole extent the declared object of his study. Each chain of arguments must end somewhere, and not everything can be examined all at once; at any given moment a great deal has to be viewed as sufficiently justified. Of course, these presuppositions might come under severe criticism at some future time; but when this occurs, there will be something else for scientists to fall back on in their changed historical situation. Nothing really ever happens *ab ovo*. Those who believed that the contrary was true, for instance, Descartes and Dingler, were deceived. The same is true for those who think that each precondition one uses can be arbitrarily set. This is just as impossible as trying to prove everything.

The permanent state of restlessness which belongs to science, its compulsion to constantly expand its knowledge, is in part caused by the problem of justification which can never be absolutely solved. This problem of justification will always admit only of provisional, ad hoc, hypothetical solutions that are bound up with a certain situation. Therefore the unsettled state of science will never come to an end. However, there is yet another reason for this restlessness as far as the historical sciences are concerned; hence there are further justifications for a priori principles and for their constant change, justifications which arise in a special way. It is with this that I now wish to deal.

13.14 The Past as a Function of the Present

I will begin with the so-called narrative sentences first treated, I believe, by Danto.[25] Among other passages, he cites the following example, taken from a poem by Yeats, in order to illustrate the nature of narrative sentences. The poem tells the story of the rape of Leda by Zeus.

> A shudder in the loins engenders there
> The broken wall, the burning roof and tower
> And Agamemnon dead.

Danto wants to show that this statement could not have been made by an eyewitness to the event—that is, provided the event really took place— 346 because no one could have known what would happen in the future. Hence narrative sentences in fact show us that things happening right now will very often look different to the historian and very often are actually something quite different when seen by the historian later on with a knowledge of intervening events. And this would still apply, even in a case where we did not subsequently obtain any further knowledge of the events reported by the eyewitness. Some things which seem very important to the eyewitness may appear negligible in the light of later events, and vice versa. To him certain things might apear to be closely related, things which later on turn out to be rather disparate or disjoint. He may describe something as a great evil which today we recognize was good. He may interpret some facts using constructions of historical systems which we now construe differently.

Let me once again stress that all this can happen even if we know everything regarding the events the eyewitness has seen, or, to put it another way, even if we have learned no new facts about the events which were unknown to him. The reason for this lies in the fact that with the growing distance in time these events will be seen in terms of different relations, in terms of relations with other events, with more and later events. We can compare this with the changing aspects of a picture or painting which come to view when at first we see it very close up and then step back from its gradually. The details of the object will show an increasing number of different connections; consequently the meaning of these details, their function, and even their content will change. Here one might also call to mind that even in everyday life we often say, "I see 347 what happened to me earlier in a different light today," and that this does not necessarily mean that we have obtained more information about what actually happened earlier. It can also mean that things simply look different if we know what resulted from them and what happened later on.[26]

I shall now give a further example which I hope will demonstrate this with the utmost clarity while at the same time serving to further illustrate and clarify Danto's insight. This is drawn from Wolfgang Schadewaldt's work, *Die Geschichtsschreibung des Thukydides (The Historical Writings of Thucydides).*[27] The discussion this work stimulated among historians and philologists concerning the details of scholarly research on Thucydides is not of importance to us here; what is of importance, however, is that Thucydides rewrote his account of the Peloponnesian War after the 348 final defeat of Athens. On this point almost everyone seems to be in agreement with Schadewaldt and his great predecessor in this field, Eduard Schwartz.[28] But then the question is: What did Thucydides write during the war, and what did he write afterward about the same events? How did these events look to him while they were occurring—that is, when he was something of a witness, or even an eyewitness—and how did they look after everything was over? In any case the answer is that they ap-peared quite different to him; and this holds regardless of the differing opinions held by historians as to details. Schadewaldt writes:

> The significance lent to the depiction of the Sicilian Expedition throughout the work, judging by its extent and formality, must rest on the significance which the event of the Sicilian Expedition had for the entire Peloponnesian War in Thucydides' own mind.[29]

From this, Schadewaldt concludes that books 6 and 7 of Thucydides' work (which treat the events under discussion) cannot have been written before the end of the war. Only with the aid of hindsight could Thucydides have considered the destruction of the Athenian army in the year 413 B.C. to have been the decisive turning point in a series of events ending with the eventual fall of Athens in the year 404 B.C. He could not have predicted this as an observer at the time of the Sicilian catastrophe; to the contrary, the position of Athens was actually greatly improved following the victory at Cyzikos in 410 B.C.

349   Only knowledge of the outcome, only a comprehensive hindsight of the final defeat, the completed "destiny," enabled Thucydides to see the forces moving behind the immediately visible setting and accordingly to interpret what happened in a new way—to discover connections where none had seemed to exist, to interpret particular events as causes, reasons, or motives for later events that could not have been initially predicted, etc.

"This work is not composed of simple reporting of plain events (*Erga*)," continues Schadewaldt, "but rather, a single, large, cohesive event (*Ergon*) (7, 87, 5) is understood historically here as a part of the 27-year war, which is now conceived as a totality of effects. The meaning of this event is recognized in terms of this universal totality; in the same manner it is structured according to its meaning, and in this way its immanent sense (*Sinn*), the sense of realty, is revealed."[30]

Accordingly, Thucydides also had to rewrite the speeches of Pericles ("Pericles speaks with the knowledge and intent of the historian after the year 404, not of the historian of 429").[31] He had to demonstrate the discrepancy between Pericles' ideals and the real forces which could not be fully recognized until later. Thucydides had to connect Alcibiades' behavior and the Athenians' aversion to him, on the one hand, with the defeats of 413, and later of 404, on the other; the connection between the two was not apparent until Thucydides was finally able to interpret the outbreak of conflicts between Alcibiades and the Athenians as a mere

350   symptom of a deeper political infirmity, as a kind of deterioration that gradually clouded the vision of his countrymen and later led to the final downfall of Athens. He had to see the entire period as *one* war and organize the comparatively long spans of peace into one cohesive devel-

opment which welded these together with the military operations. He had to valorize Athens's uniqueness and significance, something which could only have been appreciated once the period in which the Athenians had enjoyed sweeping successes and achievements in all areas of their efforts was past, that is, only after they had proved their steadfastness and intelligence in desperate and catastrophic situations. Furthermore, only in retrospect could Thucydides have used Themistocles and Pausanias as great examples of the problems of Athenian democracy and Spartan oligarchy; only in retrospect could the causes and motives for the Peloponnesian War be traced all the way back to the time directly following the Persian Wars. Finally, let me mention the so-called archaeology in the first book, which is thoroughly conditioned by Thucydides' own temporal position in history. Here he reminds the reader that at first the Hellenic tribes did not call themselves Hellenes, since they were not originally aware of their unity, even though they already had the appearance of a nation at that time, etc., etc.

Hence, in the first place, we see that it is only from a later vantage point that Thucydides comes to consider certain events as symptoms of political infirmities, infirmities that could not have been diagnosed concurrently with the events. Thus he reinterpreted these events as symptoms. When someone says that an event was the sign of a sickness, not having been able to know this in advance, he obviously gives this event a new sense. And when this event results in severe consequences, which he again could not have known in advance, the entire incident gains yet another new meaning, which could not have been attributed to it at first. Moreover, we can also relate events that formerly seemed disconnected in this manner. For if we start with the result, the sickness, then we can relate all the symptoms within one interconnected structure as symptoms of one and the same progressive sickness. In the second place, we can see that Thucydides was evidently attempting to describe what I have called a "historical system," for instance, the system of Athenian democracy, of Spartan oligarchy, of the political ideas of Pericles, etc. He viewed the sickness, which to his mind worked like a poison by gradually destroying everything, as an essential component of these systems. Thus even his description of these symptoms was transformed after the sickness was diagnosed. After all, no system is unveiled and revealed immediately.[32] It can be developed only in the course of time (explication), thereby revealing its potentials, its solid core, its basic ideas, and its contradictions, etc.; nor can the originality, uniqueness, or greatness of a historical phenomenon be known before it has run its full course.

The sickness Thucydides speaks of is the total *disorder* (ταραχή), the spiritual confusion, of the Greeks at that time. Following the loss of the original Homeric harmony, ideality and reality began increasingly to diverge. The natural will to power was no longer held in check by com-

351

352

prehensive cultural structures and ideas, but rather degenerated into an idiotic war of all against all. However, according to Thucydides this only revealed the deeper structural shortcomings of Athenian democracy on the one hand and Spartan oligarchy on the other. Athens could not avert growing demagoguery, corruption, and anarchy, while the Spartan oligarchy was at the end only trying to preserve its own power as it became increasingly ossified and impotent. In the course of Thucydides' explication, then, both systems emerge in many respects as self-contradictory and doomed to failure.

Now, even if it seems to be a fact that Thucydides did not recognize all this until he was able to view the events in retrospect, without at the same time having gathered any more knowledge regarding the earlier events as such—this being the pivotal point here—why could we not imagine a person of great prophetic powers who could have predicted the final end of the war, and who would thus concurrently have been able to write the history that Thucydides could only write from a subsequent vantage point? In such a case it would not have been necessary to change the interpretation of past events.

Here, however, one thing remains certain: In the absence of such a prophetic gift or in the absence of pure coincidence, correct prediction becomes impossible in history. No one could justify this kind of predicting in a rational manner, as there are no strict laws, not even indeterministic ones, which could be used in such a justification.[33]

353　　We are not concerned here, however, with irrational or even miraculous prophecies, even though they are conceivable and have probably occurred from time to time, but rather with gaining an understanding of how subsequent events necessarily alter the interpretation of earlier ones whenever the historian proceeds in his task scientifically, that is, whenever he restricts the manner in which he sees and describes things so that he only proceeds in a manner which he can defend as an historian. Here a brief comparison might help: A doctor can occasionally form a preliminary prognosis concerning his patient's illness without being able to prove his ideas immediately. But in this case he acts responsibly as a doctor only if he proceeds with caution and waits for the development or appearance of further symptoms so that he might be able to diagnose the illness and possibly reinterpret earlier events in an accountable or provable manner.

In order to avoid misunderstandings, I must emphasize here that I certainly am not denying the existence of so-called immutable facts, facts like the defeat of Athens in 413 B.C., the outcome of the various battles which took place during the Peloponnesian War, etc. But we find facts like these in simple or straightforward chronicles, and these should not be confused with the works written by historians. Thus we can distinguish *core facts* from those which are more or less subject to change in the manner in which they appear. For instance, Pericles' death from the plague

　　　　　　　　　　　　　　　　　　　　　　　　　　　　CHAPTER 13

is such a core fact (which does not mean that this fact is absolutely true), whereas his speeches as reported by Thucydides are not.

I hope that this sufficiently clarifies what I mean when I say that the past is necessarily a function of each respective present, contrary to widespread opinion. Nothing could be less true than Schiller's famous saying that "the past stands eternally still" ("Ewig still steht die Vergangenheit").[34] Thus the main task of the historian is not to determine "how it actually was" (Ranke—"Wie es eigentlich gewesen"), if this means how an eyewitness would have seen it. On the contrary, such a task would tempt us into chasing after the phantom of an eternal truth about the past supposedly concealed behind the curtain of the "spirit of the times" ("Zeitgeist"). The historian's main task must rather consist in the continuous rewriting of history by reckoning with the inevitable change that governs the past itself in the course of time.

How this happens is also rather impressively shown by the example of the history of the historical writings concerned with the fall of Rome. We will discuss this here briefly in order to complement the example of Thucydides.[35]

The fall of Rome is a veritable *topos* in the course of Western history which makes it possible for us to discover the vast changes that took place in the manner in which the historian conceived of himself and his time, and in the understanding governing his manner of retrospection. When Augustine considers the fall of ancient Rome as the payment for a long and excessive period of sinning, he can only do this because he is already living in a Christian time and already recognizes the end. However, from a medieval point of view, Otto von Freising sees this event not so much as the decay of a whole epoch, but rather as a symptom of that "translatio imperii ad francos et teutonicos" in which the universalism of Rome was preserved, but gradually displaced in the direction of the Holy Roman Empire. Only in this way, within such a continuity, could the Catholic idea have developed and come to fruition. The decline of the worldly Roman Empire on the one hand thus corresponds to the rise of the transcendent empire of Christ on the other. What seems to us to be a breakdown and disintegration of an old world and the beginning of a new one, hence the beginning of something separate or disjoint, appeared in the mind of Otto von Freising to be something like the connected harmonic unity of resonating pipes. If we then turn to Machiavelli, we find that he had a totally different opinion about this. Since Christian policy had gradually proved itself to be basically questionable, it was no longer possible for him to think of Rome and the Middle Ages as a unity, and he could no longer interpret the fall of Rome merely as the prelude to the drama of redemption. The standards for assessing the situation become reversed, and what was once connected is separated again. The final result of this is that the fall of the Eternal City leads to the disaster of the present (Machiavelli's present); it characterizes the ruin of an old world full of greatness, opening the way for a new world filled with misery. Hence the causes for this event could no longer be thought of in terms of a transcendent realm, since now they were to be found in the natural forces and powers belonging to men and in the contradictory nature of the principles present in their systems. We see Gibbon progress even further along this path. To the Machiavellian axioms he merely adds an enormous amount of historical material; in particular he was able to base his opinion on the infinitely more profound criticism of Christian doctrine and policy which had been developed through the rise of rationalism.

What we thus find mirrored in the works of Augustine, Otto von Freising, Machiavelli, and Gibbon are the basic structures of thought in terms of which the people of each of these time periods looked back on the fall of Rome. Here I cannot take the time to go into all the other people who, to a greater or lesser extent, described this great event in history in the same basic ways given above, differing with these authors only in the interpretation of some of the details. Again, however, we can see how it is that the historical object forces us, in the course of various historical developments, to change our opinion about what should be considered important and unimportant in it—what should be viewed as separate or connected, bad or good, etc. We can see how the alteration of the historical object changes our more precise manner of construing it and our attempt to interpret its particular details. When I say "the object changes or is altered," this merely means that the object enters into new relations with subsequent events. This is again the case because, as I have tried to show, the object is not a kind of atom to which one might simply add other atoms. To the contrary, the same object appears in a new light because it is viewed in terms of a changed constellation; and this then provides the historian with new possible interpretations to select from.

13.15 Three Forms of the Justification of Theoretical Principles in the Historical Sciences

To return to the question of the justification of a priori principles in the historical sciences, let me first summarize what has been established thus far. First: The justification proceeds on the basis of the interrelated manifold of systems within which the scientist lives. Second: The change in these principles is justified by the change which necessarily takes place within this manifold of systems for the reasons already presented (the self-movement of system-ensembles). Third: In the case of historical objects, the change in a priori principles is further justified by the necessary change in the objects themselves. To be more precise, this means that these changed principles can be justified on the one hand by the new potential for interpretation, a potential which is provided by the objects themselves, and on the other hand they can be justified through the argument that they represent a selection from this potential for interpretation which is in harmony with the given manifold of systems.

One might object here that there is something contradictory about considering changes in a priori structures to be dependent as well upon objects, and therefore determined by experience as well as by the a priori itself. However, let me again refer to what I tried to show through the use of the model concerned with the correlation between the a priori and the a posteriori. There is of course no direct determination of the a priori element by experience, that is, nothing to which the a priori as such could ever be subjected. To the contrary, any determination through experience is only possible by the introduction of conditions provided by an S-set. For this reason we can always choose whether we will allow experiences to pass judgments on theories or whether we will reject these experiences with the aid of the a priori organon (*apriorischen Instrumentariums*) which is inevitably involved here, a rejection which we could justify, for example, on the basis of that comprehensive interrelation of the manifold of systems within which this a priori organon is embedded. But basically it is always the S-set which has the last word, because it is this which defines the meaning of "experience" and "facts" in the first place. Whatever this S-set might actually look like, in the case of a historical object it will have to contain the a priori conditions that would make possible the experience of this object as something changing. Hence the historical object will repeatedly break up elements of that a priori organon for analysis provided by the S-set.

Quite often doubt has been expressed as to whether there is any sense at all in the writing of history, since this does not enable us to understand the events as they really took place, but only as we have interpreted them and viewed them in the light of our own particular times. From this 358 perspective the writing of history is nothing but a long novel in which each particular epoch mirrors itself. And indeed there is no scientific-historical truth beyond, behind, or for that matter without interpretations. There is no eternal truth, nor is there a truth which might, for example, show us "how things really were" (*Ranke*). There is, however, that kind of truth and that kind of historical experience which is formulated on the basis of the ever-changing, yet ever-newly-justified complex of a priori principles that characterizes a particular time. Each generation has to understand its past and its present in its own way; since the one cannot be done without the other, it is inevitably necessary for each generation to write history and to write it anew over and over again. The present arguments have been intended to show how this can be accomplished in a well-founded and justified manner.

PART 3

THE SCIENTIFIC-TECHNO-LOGICAL WORLD AND THE MYTHICAL WORLD

14

THE WORLD OF SCIENTIFIC TECHNOLOGY

The first chapter of this book dealt with the close connection which exists 361 between the question of foundation in the natural sciences, the numinous realm, and the realm of art. The question of foundation in the natural sciences was then expanded to encompass that in the historical sciences as well. But who would deny that there exists an indissoluble connection between what can be called "scientific reason" and what might correspondingly be termed "technological reason"? It would be impossible to gain a sufficient understanding of our historical situation without simultaneously introducing technology into our critical investigation. Only then can we return to our point of departure in the first chapter; and in doing so, we will again be able to ask, though now in a new form, the old question: What is the status of the foundation of the numinous and of art in our own world, dominated as it is by science and technology?

In general the word "technology" (*Technik*) is used in connection with such things as artificially produced devices, machines, production processes, the harnessing of natural forces to human ends, etc. In this sense of the word there is technology as long as cultures exist. Nevertheless technology has undergone such radical changes in its underlying intentions, in its self-conception,[a] and thus also in its particular aims and goals that it is more appropriate to introduce the treatment of technology via a consideration of its history than to begin, as is so often done, by at- 362 tempting to give an answer to the question, What is technology? and thereby to come up with a definition that would be suitably fitted to all times.

a. Here and in what follows the term *Selbstverständnis* (self-understanding, self-image, or as translated here *self-conception*) is applied to technology. In using such a term, Hübner is obviously personifying technology. In English we would not generally speak of technology as something which has its own "self-conception," but rather as somehow engendering a "self-conception" in the people of a given age. However, this use of personification is not intended to be *merely* figurative. It lays emphasis on the fact that technology (like system-ensembles, cultures, or science) is something which transcends to a certain extent the life of any one individual or group of individuals; it has "needs" of its own and places demands on any culture for which it becomes a determinant factor. The status of such entities (systems, cultures, and the like) can be seen as a basic point of contention between the general Anglo-American point of view and that of the German philosophic tradition. Therefore, rather than merely obscuring this problem by "translating it away" or writing it off to figurative usage or linguistic convention, it seems more appropriate to call attention to it, and thus to allow it to stand out and be recognized for the problem it is.

14.1 On the History of Technology

Two events have been particularly decisive in determining the history of technology: the spread of Christianity and the rise of the exact natural sciences.

The significance of Christianity for technology resided, among other things, in the fact that it contributed to the elimination of the slave-based economy of antiquity, and thus to the need for replacing the human labor-pool, once so vast and cheap but beginning to fail, through the use of natural energy. With the development of the harness, for the first time mankind learned how to use animals for work more efficiently; an expanding knowledge of millworks led to increased exploitation of wind and water power; the development of the art of metalworking was fostered, and with this the use of gunpowder, as well as the art of printing, were made possible.

The exploitation of natural energy on a grand scale to meet the demands of a world changed by Christianity had the result of granting technology a life of its own, even if at first this was only true in a limited sense. In antiquity, by contrast, technology had always been almost completely in the service of the state, the culture, or art, etc. Tasks and goals were set for it in terms of these various areas, and for the most part it merely sought to make improvements within the confines of the limits set. Inventions and projects that were freely developed outside these limits, as for instance those ventured by Ktesibios (third century B.C.) or Heron of Alexandria (first century B.C.), were generally looked upon as mere amusements.[1]

363

In the Middle Ages, however, despite the fact that technology was no longer principally controlled by the culture and its tradition and even though it did in fact manage to eke out a place for ground-breaking and revolutionary discoveries, nevertheless it still remained essentially limited in that it was not supported by science, and thus lacked theoretical force and penetration. Whereas in antiquity science had been chiefly concerned with the "reasons or causes of being," in the Middle Ages it consumed and laid waste to itself in the conflict with theology. Technology at this time, just as before, was contemptuously relegated to the world of the craftsman. The first real break in this pattern occurred with the rise of the exact natural sciences in the Renaissance and with the gradual melding of science and technology into that unbreakable unity under whose banner the further development of technology right down to the present has taken place. Accordingly, at the outset one was forced to begin with the simplest of matters. Thus, for instance, in 1564 Niccolo Tartaglia calculated for the first time the angle at which the barrel of a canon had to be set in order to obtain the desired range of fire.[2]

364

Here we already find exact natural science connected with a technical-practical mastery of existence. Indeed, the advances of science were always accompanied by the appearance of a technical device—the clock, the pendulum, and the telescope, just to mention a few. Along with this we find an ever-increasing demand to define the concepts of the natural sciences by means of operations utilizing measurement instruments, which were at the same time becoming increasingly more complex and encompassing in their own right. For its own part, however, technology did not merely utilize the knowledge generated by the natural sciences; rather it also brought new phenomena to light which in turn set new tasks for the natural scientist. For example: In 1824 Carnot developed a theory of the steam engine, that is, he only developed the theory after the steam engine had already proved itself to be remarkably useful for some time; in 1912 von Laue recognized X-rays for what they were, again at a time when their use was already quite widespread.

The decisive feature of this melding of natural science and technology which began in the Renaissance lies in the fact that here for the first time the practical subjugation of nature both enters into and is carried forward on the *theoretical* plane. As a result the sphere of development properly belonging to technology is no longer limited to a mere state of readiness to carry out assigned tasks, a state of expecting and preparing against possible occurrences, as was still the case in the Middle Ages. From the Renaissance onward, technology thus attains the necessary prerequisites, as well as being driven by the desire, to *systematically* investigate in an uninhibited and unlimited manner the entire scope of technological possibilities. For by incorporating scientific theory, it incorporates that element which frees itself as such from the particular and the individual and pushes forward into the sphere of the universal, thus developing its object domain in accordance with fundamental principles (in the sense given in chapter 8), and accordingly moving forward by systematizing and classifying. Hence, wherever the whenever scientific theory penetrates and takes hold of research, there and then for the first time does that research become free.

Correspondingly, we also find here the evolution of a new type of man, ³⁶⁵ a type which had never been seen before—the *inventor.* He is educated in natural science, and to this extent theoretically trained. He is concerned with systematic discovery and invention in general and thus less concerned with any particular object or topic. Economic, social, and political interests are not of decisive importance to him; often these are little more than a pretense or shelter for him to hide behind. Nevertheless he is ruled by the desire to see his projects turned to practical use, indeed sometimes to the extent that he wishes to force them upon his environment and fellow creatures, animals and men alike. We find this constitution mirrored in all the great inventors from Leonardo da Vinci to Papin, Huygens, Watt, Trevithick, Niepce, Daguerre, Nobel, Edison, etc., right on into our own times where, as a rule, the work of the individual inventor has been replaced by that of the team.[3]

Modern technology, especially contemporary technology, is distinguished from that of antiquity and the Middle Ages in that it possesses a completely altered self-conception. With the appearance of the exact natural sciences, technology began to set itself new tasks and to do so in an unrestricted manner. It awakened then, and continues to awaken and call forth a world of intrinsic needs, its own "necessities of life" as it were, needs which had never even been dreamed of previously. Its desire is directed to the methodical investigation of the infinite realm of technological possibilities. It wants to explore step by step that which is as yet uninvestigated and to try out that which is new. The spirit of technology belonging to earlier times (prior to the Renaissance) shows no sign of this. Of course, technology is still bound up today in many ways with tasks which are set for it by the state, society, the economic world, etc.; but what is essentially new, what basically controls and molds modern technology itself, is its dynamic force, the creative power which has been set free.

14.2 Cybernetics as Modern Technology par Excellence

366   This freedom finds its purest expression in cybernetics, that field which presents a universal conceptual system for the description of technical structures and procedures.

One of the most important fundamental concepts of cybernetics is the *transmission system (Uebertragungssystem)*.[b] A transmission system is understood as the transition, governed by operators, of certain entities (those of the input) into certain other entities (those of the output). These entities are known as operands. A simple example of this can be found in the piano. Pressing on particular keys results in the production of particular tones. Transmission systems can be developed along various lines. Thus sometimes each operand can only be transformed into a single different operand, sometimes into several others; the operands might present a continuum or they might be discrete; the transformation can be deterministic or statistical in nature, etc. In general, transmission systems can be both mathematical models and exact theories, etc. (as in physics), as well as real occurrences which exhibit ordered chains of events. Of particular significance here, however, are real occurrences in which the operands of the input or the operators can be altered, as for instance by the presence of certain modes of intervention, the actions of levers or gears, or something similar. In such cases we speak of *control, regulative,*

b. An *Uebertragungssystem* (transmission system) might also be thought of as a system for the "transference," "transition," and/or "transformation" of one thing into another. As such these other terms will also come into play in the general discussion concerning this concept.

and *matching* processes,[c] in general of any kind of *feedback processes*. These can be collectively called "transmission systems with input" (*Uebertragungssysteme mit Eingabe*).

Now, in every technical production process a preestablished goal or aim is realized in a physical-chemical manner. Thus we achieve the desired goal by means of laws which determine the transformations leading from the initial quantities, or input, to the resultant quantities, or output. Consequently, such a production process is always a transmission system; moreover it is always a feedback process, since the nonclosed nature of the physical-chemical systems involved in the production process requires that these systems be kept under constant surveillance and control. In addition, the various functions of the products thus produced also have to be regarded as transmission systems, something which comes to light when we survey the purposes behind the production of these products. These purposes can be divided into three categories: the *conservation of states,* the *use of energy,* and the *acquisition of information*. With regard to the conservation of states, the input factors are the modes of intervention by which states can be altered, whereas the resultant output factors are those states which have to be maintained or conserved. (Examples would be things like dikes, storage bins, preservatives, and heating systems.) In the use of energy we are dealing with a regulative process of transformation, namely, that of energy, and thus with a transmission system. (Here one might think of automobiles, airplanes, trains, etc.) Finally in the acquisition of information, words (messages) are fed in, transferred to electromagnetic waves, printed letters, punch cards, and the like, and then read out again at the end as words (messages). (The computer serves as an example of this.)

Hence, because in all cases production processes and their products exhibit "transmission systems with or without input," control, regulative, and matching processes, etc., they can be described by *mathematical models* whenever the laws or rules governing their transformations admit of exact formulation. And this is the case since by using such models all we are doing is ordering these laws and rules into an axiomatic format. (An example of a transmission system which utilizes rules as well as laws is the computer. Concerning the distinction between "laws" and "rules,"

367

368

c. The German terms here are *Steuerungs-, Regelungs-, und Adaptionsprozessen*. Literally translated these would read, "steering, regulative, and adaptive processes." However, these notions have picked up highly specialized meanings in regard to various fields or branches of technology. The translation given here, and largely suggested by Professor Hübner himself—control, regulative, and matching processes—cannot hope to be adequate to the various forms, and as yet still highly fractured jargon, of technology. It is in fact very difficult to find *any* standard terms, let alone cross-cultural or translinguistic consensus concerning such terms and their translations, when dealing with a field as current and unsettled as cybernetics.

see chapter 8, section 8.1.) The construction of such models has a considerable and rather far-reaching significance, since they serve as the foundation for three levels of theoretical consideration which become increasingly abstract. On the *first level,* by utilizing the model and disregarding the immediate purposes and particular forms of the technological objects involved, we can explore the scope and free play for further possibilities belonging to these objects. Here the model merely serves the same purpose as theory in that it allows us to present individual phenomena as derivable from it and, by arranging and classifying these phenomena within a larger systematic context, to gain a clear overview of the elements of the system. On the *second,* more abstract *level,* the structure of the transmission system formulated and depicted in terms of mathematical models is checked against possible reciprocal substitution with other transmission systems. For instance, when it is determined that an isomorphic or homomorphic relation exists between a technical and a natural transmission system—hence when there is either complete or partial structural agreement—the technical system will produce, either in full or in part, what is produced through the natural transmission system. It is only because switching algebra (*Schaltalgebra*) and propositional logic both correspond structurally with a Boolean lattice that certain logical operations can be transferred to technical devices. Finally, on the *third level,* by beginning with given transmission systems, we move via combinations and variations, etc., to the free construction of other transmission systems on a multiplicity of different planes, so that we can then examine how these might be practicaly utilized. We find these three levels of progressive abstraction and theoreticization in numerous fields which have just recently arisen: for example, in *switching theory* and the *theory of automata,* in *control or steering theory, game theory,* the *theory of matching systems (Theorie der adaptiven Systeme),* in *neuron models, structures of language, information theory,* etc.

369

Thus cybernetics is to be understood as the most extreme mode of abstract consideration belonging to technology, a mode which aims at the introduction of universal fundamental concepts and methods, the creation of mathematical models, and the investigation of the structures presented by these models. As such, cybernetics has proved itself to be an exceedingly fertile discipline and one which offers the unchained spirit of invention both a resource for help and an indispensable means of orientation in its press to expand both the scope of possible technological goals and their realization.

14.3 Society in the Technological Age

This purely theoreticized technology, first projected in the baroque period and later realized as such, is a technology which is determined and defined above all by the decisive emphasis it places on *progress* and *exactness.*

It is determined by the idea of progress insofar as its theoreticization is intended precisely as a means to free it from the immediately concrete, from a given purpose, the appointed task, and the given means (machines), so that it can investigate the full scope of technical possibilities in a systematic manner. (Accordingly, this can be understood in terms of the notions of Progress I and II given in chapter 8.) It is determined by the idea of exactness insofar as this goal is to be achieved by means of mathematical models, schematized transmission systems, etc. At least in part, then, technology is a kind of *game* that thoroughly embodies the maxim *l'art pour l'art*. In this way not only do new means for achieving old ends arise, but countless new goals and needs as well. An avalanche of progress roars down on us, and if there is to be an enduring revolution anywhere today, it will doubtless occur within the arena of technology. I repeat: Just as in earlier times, much of what is at work today in this process is not directly technological (e.g. politics and economics); but the fundamental technological orientation described above belongs to our times as something which is essentially new. This orientation now has a rather far-reaching effect in molding the forms of society which arise in conjunction with the technological age. 370

In the first place modern industrial processes, and thus a great deal of what makes up the modern work world, are essentially characterized by having an exact form. In earlier times this form of exactness was completely unknown. It was only with Maudslay's invention of the lathe support and movable slide that metal parts like cranks, axle shafts, valves, etc., could be readily utilized in industry. Previously—by way of assessing the historical development—as Nasmyth noted, "no system had been followed in proportioning the number of threads of screws to their diameter. Every bolt and nut was thus a speciality in itself."[4] Nasmyth thus 371 called the systematization, order, and system of controls which Maudslay introduced into workshops "the true philosophy of construction."[5] It was precisely this exactness (gear levers, pushbuttons, conveyor belts) that led to mass production and mass consumption, since exactness made possible the simple, quick, and easily repeatable use of uniform elements (operands) in accordance with strict rules and laws (operators). Indeed, we can even go so far as to say that this is chiefly what "the exact" consists in. This we see above all in its ideal form, that is, in every kind of calculative system (*Kalkülen aller Art*). These are not directly in the service of truth and insight, but rather they are principally directed toward schematized operations having certain fundamental forms (figures, symbols, etc.). It is always only a form and never a content which can be exact and which can make possible schematic operations. In these a maximum of intersubjectivity is attained, since, owing to their univocity and rigor, they can be fundamentally confirmed and understood by everyone alike. But it is also precisely for this reason that a schematic operation reveals itself as something rational. Thus a society for which mass production and mass consumption are the determinant factors is a society which tends constantly toward "rationality," no matter how unclear this concept might remain for most the people in it. In this we find the roots of the progressive "de-tabooing" and "de-mythologizing" which we observe everywhere today.

372   There can be no doubt that, standing shoulder to shoulder with the idea of exactness, the other idea which dominates our society today is that of progress, an idea which is inextricably bound up with the essence of modern technology. (It is because of the particular lack of clarity surrounding this idea, and because so many false presentations of it are currently in circulation, that I chose to investigate it in such detail in chapter 8 and not at some later point in the text.) Indeed it can be said that this idea was already ripe at the time of the Enlightenment. Accordingly we also find that the rationalization of the world, that final goal for which the Enlightenment strove, first appeared in the realms of the natural sciences and technology. With regard to this, little has changed. Today it is still the idea of progress which principally holds sway in both of these realms; and no matter how vaguely it might be conceived in the general consciousness, it is still this idea which constantly draws us onward and is upheld by our form of life. And just as this highly determined technological rationality has been broadened into an indeterminate concept, so too do we find that the idea of scientific progress has been universally extended to a point where it is thought to encompass almost everything. To rebel against rationality or progress today is for most people tantamount to protesting against the divine world order in times past. So it is that the way in which present-day human society, as an industrialized society, understands itself rests, to a very great extent, on genuine technological-scientific forms and ideas.

14.4 Technology: Pro and Contra

One part of the philosophic community welcomes this occurrence, while another part rejects it. For those who embrace it, technologization is above all else the foundation for a widespread freedom: Technological progress frees us from the grip of tradition; mass consumption and mass production

373   free us from material need; the intersubjectivity of work and the exact standardization of products contribute to a general breaking down of social differences; rationality ostensibly excludes the mysterious. The next step would thus be to enlist the aid of technology in the strugggle against all forms of taboos, a struggle to some extent demanded by and to some extent already accomplished by political freedoms; hence we are to align technology with modern democracy (Wendt,[6] Fink[7]). Thus many see the technological world as the "realm of autonomous individuals,"[8] and upon it base their hopes for the realization of higher values in a degree hitherto

unknown, linking it in this way to an older notion of the "regnum hominis."[9]

In progress they see the "emancipation of man from his imprisonment in the plant and animal kingdoms," and thus the means "to all things spiritual."[10] Higher tasks, they argue, can only be accomplished with the kind of independence and leisure time that technology creates. It also makes possible an ever faster flow of information, which in turn supposedly fosters education and helps individual people as well as different cultures to understand and get to know one another better. A universal humanization, reaching even to nature,[11] is thus thought to go hand in hand with technology.

In opposition to this optimistic assessment of technology, we find the following points raised: The addiction to the new and the complete severing of all traditional ties, factors which are woven into the very foundational stance of technology, together with the rapid and incessant change of the material environment resulting from technological progress—all of this transplants man into a restless world without roots, in which he loses orientation and his contemplative peace of mind. This progress consists solely in a maximum of activity with a minimum of reflection over the why and to what end.[12] The kind of spiritual-intellectual (*geistige*) orientation necessitated by technology, where the ideal is that of exactness and hence merely formal, does not allow for a binding value structure by means of which a person can direct his actions. Where the major activity is that of playing a game with univocal forms according to rules, where the if-then relation is pushed into the foreground and much less is said about content, about the gravity and meaning of the initial conditions or results, we have no foundation at all for obligatory values. As such, technology is held to be value-free (Litt,[13] Spranger[14]); and it is precisely for this reason that it is so easily misused. Because its core is rationality (Fischer[15]), the all-powerful spirit of technology, directed as it is toward the makeable and the doable, remains one-sided and above all shuns any relation to art and religion. Further, these thinkers argue that the new prosperity of man has not in fact brought with it leisure time and independence. Gains made in the expenditure of human energy and time are offset by the monstrous energy consumption inherent in gigantic industries and the constant lack of time found in the workaday world determined as it is by haste and an accelerated pace of life (Jünger[16]). In place of material need, we now find the press of an ever-increasing number of new needs, needs which, since the disappearance of material need, have nonetheless been found to exert no less pressure than the former. The equality (*Gleichheit*) generated by the intersubjectivity of technical work and the standardization of needs and products has become a mere leveling, an equality of indiscernibles (*Gleichgültigkeit*). People feel themselves to be numbers behind which their individuality disappears and their spiritual

374

375

powers atrophy; they pay for freedom by becoming "depersonalized"[17] and submerged into the masses. In countries marked by a high level of technology, freedom readily becomes the tyranny of the masses or the tyranny of demagogical leaders or soulless technocrats and bureaucrats. It is precisely technology itself which makes possible total control by the state and threatens mankind with annihilation through the amassing of enormous quantities and deadly weapons. The faster flow of information does indeed foster education; but it is an education of conformity, fostering the growth of a uniform world that is wretched and soulless. A technology which penetrates deeply into nature itself and transforms it as a whole into a tool (Spengler,[18] Scheler,[19] Heidegger[20]) not only leads to the destruction of the balance of nature and the whole ecological system, the consequences of which are incalculable, but also robs nature of its symbolic power—its ability to be viewed as an image of divine order.

Thus the "humanized" nature created by modern technology would in reality be nothing more than an expression of the inhumanness inherent in this technology.

376   Now the first thing we can state with regard to this is that the supporters and critics of technology alike base their arguments on the fundamental ideas mentioned earlier—exactness, rationality, and progress—even though it is rather obvious that these ideas are not always clearly understood by them. They only differ in terms of the conclusions they draw from these ideas. Thus we can summarize the two views as follows: Whereas those friendly to technology lay emphasis primarily on the view that exactness, rationality, and progress will lead to man's liberation from all kinds of external coercive forces and in the end to a "state of freedom," the critics see in all this the instigation of new coercive forces which will eventually lead to a world ruled by despots and above all to a world bereft of meaning. This divergence in the assessment of technology only serves to point out that both sides have seen an element of the truth.

In general we can agree with the critics in saying that it is sheer fancy to link ideas like "spirit," "cultural education," "humanity," etc., to technology by means of a vague employment of these concepts and further that it is illogical to expect technology to bring about the final realization of traditional values, when in fact it has been effectively engaged in the extensive reinterpretation, often the utter and complete destruction, of these values. On the other hand, it is equally unproductive to castigate technology dogmatically as something which leads only to the destruction of the traditional notion of humanity, and thus sullenly to oppose it on this ground. This traditional notion of humanity is itself hardly the kind of clear and distinct idea that many seem to think it is. (If it were, it would scarcely have suffered such a fate.) What is more, the beneficial effects wrought by technology in many areas are so incontestable that it would simply be absurd to wish to deny them.

However, even if we affirm that both sides—the supporters as well as the critics—are both right and wrong to a certain extent, we should never- 377 theless not allow this to lead us to the conclusion that by appropriately availing ourselves of technology it would eventually be possible to placate its critics. How often has it been stated that technology is neither good nor bad in itself, but rather that it is only a matter of putting it to the right use! But this hope is deceptive. Technology will never cease from also being a vexation—and I emphasize the *also*—since it is principally based upon the notions of rationality, exactness, and progress, and on the general stance which these necessarily entail; such structures as these cannot be made to conform with certain traditional value systems which are deeply rooted in our culture—or at best such can only be accomplished with great difficulty. Hence we can add that the gains of technology will always be had at the expense of the numinous realm and the meaning and significance of art.

14.5 Technology and Futurology (Zukunftsforschung)

In recent years the critics of technology have received new impetus from the gloomy prophecies emanating from futurology, a field which has been growing quite rapidly.[21]

This form of research is a direct result of modern technology and consists mainly in technological "forecasting." Today, as we have shown, technology is indeed essentially future-oriented. Contrary to earlier conceptual schemas, the old means nothing to it, while the new and change mean everything; a climate in which inventive genius thrives. But futurology actually begins as soon as we reach a point where the effects of this 378 dynamic process can no longer be immediately seen. Here mankind begins to take on the appearance of a necromancer who can no longer control or contain the spirits he has conjured up. Our developing technology seems to be reaching this point with alarming speed. It is feared that within the foreseeable future we will reach a stage of universal pollution and contamination of the air, water, and food supply, coupled with an immense increase in the world population, which already stands at a staggering figure. In such a world every production process, no matter how large, would be both hopelessly outstripped by demand and in part offset by side effects. To restrain, or retreat from, technology seems impossible without condeming the world to misery and poverty, while to go on in the present manner, it is asserted, will eventually bring on a catastrophe.

Obviously everything depends here on whether we can correctly predict the future and the effects of the precautionary measures we take. The field of futurology has developed various methods to this end. To mention only the most important, we find methods like *trend extrapolation,* which attempts to determine the future course of a trend on the basis of the conditions figuring into its origin; the *relevance-tree technique,* by which one hopes to predict future critical points on the basis of a quantitative evaluation of the significance of every aspect of a system by which goals are set; the *Delphi technique,* which rests on a *convergence procedure* in which the prognoses of competent specialists are weighed against one another; and the *morphological method,* by which the most favorable solution to a technical problem is sought. However, all of these methods suffer from the common weakness of having been developed on a more or less ad hoc basis, and thus of being theoretically unsatisfactory.

379   A theory of futurology has to be grounded on a theory of historical processes. Indeed, the intention of such research—namely, to predict historical processes—presupposes that one already has a concept adequate to the description of the form which such processes must follow in the course of their development and the structures they must possess. Thus it becomes readily apparent that the more technology directs itself toward the possible and the futural, the more deeply it infiltrates and takes hold of the spiritual and material roots of existence, thereby creating a world increasingly indifferent to the past, the more necessary and unavoidable this reflective consideration of the historical becomes.

14.6 Technology in the Light of the Theory of Historical System-Ensembles and the Pathos of Change

For the above reason we must ask ourselves what the history of technology has to teach us with respect to this problem and what structures the development of its processes actually exhibit. In terms of those considerations with which we began, the first thing we can say is that technology does not make possible a quicker and constantly better satisfaction of a universal substratum of human needs. To the contrary, the needs which presently lie at the basis of technology and the technological society are themselves the result of a process of historical transformation. The idea of making science "technological" and technology "scientific" (theory determined), as well as that of comprehensively altering the world so as to bring it into line with the viewpoint and framework of rationality, was completely foreign to the earlier conception of technology, a conception which was most essentially determined by handcraft and the traditional tasks which were seen with respect to the limits set by the state, the culture, etc. As I will now show, the ideas just mentioned are neither given by an eternally valid form of reason nor occasioned by pressing

380   empirical needs. Rather, what we have here is a historically conditioned phenomenon just like the appearance of Christianity or the rise of the exact natural sciences. Technological needs, when we ignore the trivial elements here, are just as mutable as all the other appearances belonging

CHAPTER 14

to a culture. Hence, in writing the history of technology, it is again impossible to remove or abstract it from its connection with the other areas of a culture (like politics, art, the economy, etc.).

The greater part of those philosophers who have concerned themselves with technology have failed to recognize its historical constitution. Thus, for example, Marx seems to have believed that since the very dawn of man technology has been undergoing an internal process of self-development, the significant stages of which were accompanied by social upheavals. Indeed, according to his conception, the driving force behind world history is the contradiction which exists between incessantly expanding productive forces on the one hand, forces which are somehow always taking on new forms, and the restrictive social structures on the other, which lamely limp along behind these new forces. The introduction of the weaver's loom he holds to be the cause of medieval feudalism; that of the steam engine, the cause of modern "civil society." Seen from this perspective, technology autonomously unfolds in every age, other factors serving at most only to restrict or foster its growth. Everything else eventually accommodates itself to technology, though of course it might at times entail a struggle.[22] Another conception of technology, equally nonhistorical in nature, is that of Friedrich Dessauer. For him man is at least always latently "*homo faber*," an "investigator" and "inventor," and hence a "technological being."[23] In this conception, man is driven on by need as well as the desire for luxury, profit, power, or spiritual fulfillment. It is supposedly for this reason that he first made a fire, built houses, cultivated the land, forged weapons, laid out streets, and built dams. By busying himself with such tasks for thousands of years, he acquired more and more practical skill and made an ever-increasing number of discoveries, which finally sweep down on us today like an avalanche, transforming the entire world into a technological order of existence. Thus, while according to Dessauer technology increasingly frees us from original human needs, we find that, as in Marx, these needs are constantly being redefined and generated anew by technology itself. But in both cases the way in which technology is understood, its very self-conception, is looked upon as unchanged, and to this extent it is viewed nonhistorically. And in this respect we find that Kapp,[24] du Bois-Reymond,[25] Mach,[26] Spengler,[27] Diesel,[28] and many others have thought along similar lines.

To the group of thinkers who have emphasized the historicality of technology belong Ortega y Gasset[29] and Heidegger.[30] Ortega distinguishes between the technology of the craftsman in previous eras, where custom and usage dominated rather than the drive toward the new, and that of the modern age, where technology turns more on an all-encompassing, unlimited activity less concerned with particulars. According to Ortega, technology has in fact always h. d a state of well-being as its aim; but the nature of this concept of human well-being is a problem which obviously

381

382

can only be dealt with within the framework of a culture which itself undergoes an historical process of coming into being and passing away. Heidegger also emphasizes the different natures belonging to modern technology and that of earlier times. Whereas once it had been a cautious "uncovering" (*Entbergen*), today it consists in a violent and forceful production (*Hervorbringen*). For Heidegger the products of modern technology are no longer preeminently objects of veneration and things to be pondered, but only "that which can be ordered" (*das Bestellbare*) and "that which is at our disposal" (*das Verfügbare*). Thus today technology represents a defiant "demanding of nature," a kind of trap (*Gestell*) in which nature, like the wild animals which belong to it, is constrained or "trapped" into doing our bidding. Even if one admits that the analyses of Ortega and Heidegger are inadequate and outdated in certain particular respects, we can nonetheless say that they grasped the historical constitution of technology quite correctly.

To understand technology historically means to grasp its history as the history of its fundamental aims and norms. It would be nonhistorical, then, if we were to think of these aims and norms as remaining constant, forming the sole fixed framework within which all technological development occurs. It is precisely these fundamental changes in technology's own self-conception—changes which can be seen in the great epochs, especially in the upheavals which mark off antiquity from the Middle Ages and the Middle Ages from modernity—which show that technology must be understood historically. If we wish to characterize the self-conception of technology belonging to any one of these periods, then we will inevitably have to do so by placing this self-conception within the organized matrix of those fundamental aims which mark that period. Thus we see, for instance, that modern technology *emphasizes* rationality, exactness, and progress, and therefore also strives to theoreticize the practical, to systematize everything within a comprehensive unity, etc. On the basis of certain fundamental orientations we can then obviously deduce others. Hence we can say that the structure of modern technology is organized as a more or less rigorous system and that it can be described by means of a corresponding system. All other historical epochs of technology will then admit of a similar analysis, if we take the trouble to trace them back to unified fundamental aims and purposes. What this means can be summarized in the following: *The history of technology can be interpreted as a history of systems and thus, in terms of chapter 8, as a history of the explications and mutations of these systems.* By way of example here, we can say that the development of modern technology can be viewed as an explication of a system whose fundamental aims, as already stated, are those of rationality, exactness, and progress. By contrast, the revolutionary transition from the technology of the Middle Ages to that of the modern era would be viewed as a system mutation, since in fact this

383

transition occurred in such a manner that the new system was not directly deducible from the older one, but rather contained new fundamental aims and purposes.

Now since system mutations and system explications are structures of historical processes, it follows that futurology must consist precisely in the prediction of such explications and mutations. Accordingly, it must deduce these future structures from existing reference systems in accordance with principles (e.g. those belonging to technology or its countless subsystems). But the unavoidable and insurmountable difficulty in doing this is to be found in the fact that on the one hand the continuance of the reference system must be presupposed, while on the other a prediction of such a continuance is impossible. The continued existence of the reference system could be possible if and only if the foundations of the system in question were necessarily valid, absolutely evident, or the like. But it is precisely this which never proves to be the case with historical systems such as those of modern technology or its various subsystems, at least not from the scientific point of view. We cannot have prophetic insight into the future of historical systems or their subsystems. But why is this the case? Precisely because, as we have already stated on numerous occasions, there are no necessary contents belonging to an eternal form of reason upon which to base such foundations and because it would be just as impossible to refer these back to necessarily binding experiences (cf. chapters 4 and 8). 384

We can make this even clearer in the present context. Where is that experience which is supposed to compel us to pursue technology in the modern sense? Do we have some kind of experience which informs us that technology is better able to realize our ideas of a better, more comfortable, more beautiful, more pleasant, in every case more desirable life? Whoever affirms this must presuppose that ideas of what a desirable life is are anthropological in nature, hence deducible from an immutable human essence, and that because such eternal aims are posited and established, it follows that we can have experience of the means necessary to achieving this end. But even ideas of what a desirable life would be are considered to be historical by the historian. What is more, as we have seen, modern technology never even bothers to question the fact that technology sets aims and goals regardless of whether or not we all hold them to be desirable. In any case, the historicality of such aims shows that they are not the product of some kind of necessary and absolute system of ends pertaining to a self-positing reason. Moreover, reason has not always been linked to exactness and progress. Even if we were to ignore this and stress the rationality inherent in modern technology above all else, we still could not assert that because of its rationality technology is something which reason must necessarily desire. The rationality of technology is in and of itself something formal, as can be gathered by 385 looking at the *l'art pour l'art* of systematic invention, the schematized operations of industrial processes, or the consideration of cybernetics as a kind of fundamental science operative in it, directed as it is toward naked formal structures. And that which is distinctively new in present-day technology, as the previous analysis clearly shows, is principally its tendency to place this formal element in the *foreground;* for it is by means of this that it ultimately triumphs over all those problems of content which, unlike in former times, are reduced here to a rather low level of importance. However, this kind of *absolutization* of the rational does *not follow directly* from rationality itself, since it rests on a claim that is made *about* rationality. At most it can only be rationally founded in an *indirect* or *mediate* manner, namely, by again deducing it from certain *givens* of one kind or another, which cannot as such be rational, and then starting from these. But since these *givens,* as already shown, also fail to present us with any kind of "absolute experience," they can only be historical as well. So technology, insofar as it aims *primarily* at rationality, exactness, and progress, as if these were its own essential end, can itself have nothing other than historical roots and cannot be derived in some absolute manner from reason (*Vernunft*) understood as rationality (*Rationalität*).

We can now see that what I said earlier—namely, that recourse to the consideration of the historical becomes all the more urgent as technology and the world determined by it, with its almost total orientation to the futural and the possible, becomes increasingly indifferent to the past— becomes even more significant, since within the present context this recourse to the historical takes on a meaning which is yet deeper and even more universal. Initially I made use of this statement by way of referring primarily to the preeminent concern over the future which has developed in conjunction with the accelerated growth of technology; but now we also see that the historical element of technology has become increasingly apparent *precisely because* technology has itself become objectified rationality. Because this unleashed rationality, indifferent to content, is necessarily afflicted with the pathos of change, it is condemned to change— condemned to a permanent technological revolution, to a continuous "trying out" of possibilities; and for this very reason it is condemned to an incessant overthrow of that reference system to which all expectations and all calculations might finally be related in technology's attempt to master itself. In the end a will reveals itself within this pathos, a will to consume and lay waste to itself in its own historicality. This orientation toward rationality has been at work throughout history whenever attempts were made at harmonizing systems, simply because this harmonization is itself something formal. But previously it had always been limited to those transitional periods in which this harmonization took place. Hence this rational orientation was less of an end in itself and more of a means to other ends: Before as well as after the transition, the primary interest lay in the consideration of contents which were *sub specie aeternitatis*. It is only with the age of technology, and even here principally within the realm of technology proper, that this orientation toward rationality comes to be something of an end in itself, in any case becoming a fundamental orientation. No one believes in fixed contents any longer, but rather only in "models." In fact today everything is considered merely as a "model"— a favorite word of our times, which is employed in every aspect of technology—and a model is never thought of as something final and exhaustive. Hence the more technology revels in its rationality and its formalism, the more unpredictable and unmanageable it becomes, and consequently the more difficult it becomes to control its future course by means of scientific forecasting.

Let me summarize: In the light of a historical consideration of technology, cybernetics proves to be technology par excellence; that is, it shows itself to be the most highly theoreticized form of technology, thus that form directed to the whole universe of practical possibilities in general, as a totally rationalized technology directed completely at the future, at progress and change. This yields the fundamental appearances and relational forms belonging to an industrialized society, as well as all the factors, both pro and contra, which relate to such a society. But precisely because this society is principally directed toward the possible and the futural—an orientation which is completely different from that of earlier forms of society, bound as these were by "Being in its eternal present" and the past as a depth of tradition (a point dealt with quite extensively in the last chapter)—we find that concern over the future comes increasingly to the forefront. Technologically oriented as our world is, it also desires to grasp the future technologically. But it can only do this to the extent that it employs the structure of historical processes and insofar as it understands technology itself as historical process. Thus, having almost become lost in the future, we are emphatically reminded of history. But it is precisely in this manner that we find revealed to us the nonrational conditions of our illusory image of absolute rationality. We see that the more we abandon ourselves to rationality, the more deeply entangled we become in our historicality. Our ability to master the future becomes more problematic as the world becomes more technologized.

What I have attempted to give here is an analysis, a diagnosis, if you will, of the technological world. In doing this I have not shied away from giving a definition of the essence of modern technology. But this essential definition has nothing in common with that kind of essentialism that is constantly and legitimately criticized, since here it rests on a limited historical phenomenon (just as there is a difference between asking what prime numbers are and asking what numbers in general are). An analysis, diagnosis, or definition of the essence of technology has, among other things, as its primary purpose the task of clarifying the underpinnings of

387

388

the situation in which we find ourselves. The other, and more frightening, question of what is to be done in this situation, a question of such pressing urgency today, remains unanswered here. But how can we give any kind of satisfactory answer to this if we do not first understand precisely what the foundations of our situation are? I believe that all those who cry out for a change in this situation, developing more or less useful ideas concerning both how this might be effected and the goals which are to be sought, are still for the most part, even when we give credence to the legitimacy of their concerns and work, merely busying themselves with palliatives. Whether or not we should pursue capitalism or socialism, whether this or that form of society is best, or whether this or that technological, economic, or social plan should be implemented, are indeed extremely important matters from a political point of view; but *philosophically* viewed this is not the *fundamental* problem. This fundamental problem is the same everywhere, in the East as well as the West, a problem which has to do with the effective self-conception of modern man—that is, with his technological-scientific form of intentionality, and thus with the emphasis placed on rationality, which today has nearly succeeded in becoming an end in itself. Herein lies the essential greatness of man today, as well as his essential weakness.

Excursus concerning the Theory of Rational Decision Making

After the previous remarks, it is easy to understand why rational planning is one of the key concepts of our times. There are several theories in circulation today which go by the name of theories of rational decision making (and planning obviously means decision making); here I have in mind the theories of von Neumann, von Morgenstern, Bayes, Ramsey, etc. Since I cannot discuss all of these here, I will limit myself to the analysis of a model which, though rather simple, seems nevertheless to be representative of these theories, since we find the same fundamental problems present in it as in these various theories.

Let us suppose that Mr. X wishes to reach a particular goal and there are different possibilities open to him by means of which to do this; further, that for each of these possibilities there exists a multiplicity of significant circumstances. One of these possibilities might quickly bring him closer to his goal under certain circumstances, while under other circumstances he might be forced into detours and a considerable loss of time. Accordingly, Mr. X's first step would be to project a matrix in which the possibilities form the lines, the circumstances form the columns, and the respective results of the possibilities and circumstances make up the elements. Such a matrix is called a *matrix of consequences.*

389

$$
\begin{array}{c|ccc}
 & C_1 \cdots\cdots\cdots\cdots\cdots C_m \\
\hline
P_1 & R_{11} \cdots\cdots\cdots\cdots\cdots R_{1m} \\
\;\cdot & \cdots\cdots\cdots\cdots\cdots\cdots \\
\;\cdot & \cdots\cdots\cdots\cdots\cdots\cdots \\
\;\cdot & \cdots\cdots\cdots\cdots\cdots\cdots \\
P_n & R_{n1} \cdots\cdots\cdots\cdots\cdots R_{nm}
\end{array}
$$

In a second step X will assign a particular use-value to each element, hence to all the results presented in the matrix. This use-value tells him how useful the respective result is in terms of his original intention of quickly reaching the goal. In this way the *use matrix* is generated from the matrix of consequences.

In a third and final step, X determines the probabilities governing the occurrence of the respective circumstances; and in this way he sets up a *probability matrix*. Thus by virtue of this model he now has everything he needs in order to choose rationally between the various possibilities at his disposal for reaching the goal; this amounts to saying, according to a rule introduced by Bayes, that he will calculate the expectation value for every possibility and then determine his action in accordance with that which offers the highest expectation value. This value is given by the formula

$$
\text{Exp.}(P_i) = \sum_{k=1}^{m} U_{ik} Pr_{ik} \, ,
$$

where U_{ik} is the use-value of the action A_i under the circumstance C_k, and Pr_{ik} stands for the probability governing the occurrence of circumstance C_k. (For the sake of brevity, the presuppositions involved in the probability theory in use here have been omitted.)

So much for the model; that is to say, it is readily apparent that the notion of rationality this model presumably expresses presupposes certain intentions and assumptions on the part of Mr. X: *If* Mr. X is intent on reaching a particular goal; *if* he assumes that there are only such and such a number of possibilities for doing this, that this or that circumstance plays a role, that this or that probability can be assigned to these circumstances, etc., *then* he can go on to calculate the different expectation values and to make the rational decision as to how to proceed. However, it is obvious that our model tells us nothing about the essential nature of the rationality of all these goals and assumptions. Might these not just as readily be the goals and assumptions of the madman?

Theories of this sort are then quite obviously too weak to provide us with a satisfactory answer to the question of the rationality of decisions;

however, they can show us where we must look in order to further our investigation of these matters. Clearly we must turn to the intentions and assumptions just mentioned, and thus to the content of the matrix posited by X.

For example, what does it mean to say that Mr. X rationally arrives at the elements of the matrix of consequences, i.e. the results? These results are prognoses which rest in part on natural laws and in part on rules of human practical activity. For instance, let us say that X has it in mind to take a trip. He considers the following: If I take an airplane and run into fog at the airports, then because of existing natural laws the result will be a significant delay. Or X might want to make money on the stock exchange. He considers: If I invest in bonds, then owing to the existing economic rules, the chances of making a profit are small, etc. Therefore in this context rationality can only mean the rational foundation for those natural laws and rules which are employed in his prognoses.

Foundations and groundings of this kind are, however, highly complicated matters—something which has been quite clearly shown in the previous chapters in our consideration of the questions of verification, confirmation, the falsification of laws and rules, induction, etc.

Things become even more complicated when we move to the consideration of the elements of the probability matrix. What do foundation and grounding mean in this context?

There are scholars who deny that such a rational founding is possible at all. Others have sought to develop various theories by means of which statistical hypotheses can be rationally checked or supported. But unfortunately it can be shown (1) that these theories as well cannot be universally valid; (2) that they already presuppose ideas about how one goes about determining any and all probabilities in the first place; and (3), the most important of all, that the acceptance or rejection of a statistical hypothesis never presents us with a simple alternative and can never have the force of necessity.

Hence we must conclude that the theories of rational decision making available today presuppose theories concerning the justification of laws, rules, and probabilities—theories whose rationality is equally questionable.[31]

Here I will pass over the question of how X arrives at the formulation of the different possibilities for reaching his goal or the elements of the use matrix. I wish instead to concentrate exclusively on what appears to be the most important question of all: How can goals themselves be rationally founded?

We must never lose sight, I believe, of the fact that aims or goals are again never solitary monadic entities; rather it is always the case that they can only be given within the context of a particular situation. We always find ourselves already living within a system of private and public ends or purposes. Accordingly, it would be senseless to ask whether an individual goal in and of itself is rational; we must rather direct such questions to the entire systematic context within which a goal appears—that is, to the systematic context within which we live and move, and therefore from which there is no escape. To be sure, it is possible for us to change the 393 order of importance granted to elements within this framework or to allow one element to be assimilated by another. But precisely because we can never completely leave behind that system-context which we are locked into—since "life" always means "life within such a holistic structure"— the rationality of this system-context will always consist, I maintain, in the fact that the system-context is harmonized in the sense attributed this term in chapter 8. The degree of its rationality will consequently be dependent upon the level of harmonization which has been attained. Thus we will have to judge a goal in accordance with the extent to which it fits into a given comprehensive context and by assessing whether it contributes to the elimination of inconsistencies belonging to this system-context. By way of contrast, we might then say that it is the height of pure idiosyncrasy to think in a manner which completely neglects that mode of consideration which views things within the context of the whole—an idiosyncrasy which might perhaps be accounted the essence of insanity when the latter is taken as that state most completely opposed to rationality. But in conjunction with this I would like to reiterate what I said earlier in chapter 8, namely, that in certain situations rationality can also contain contradictory and divergent goals; but this can only be the case if the appearance of such contradictory and divergent goals, while unavoidable from time to time within a limited context, nevertheless still contributes in the long run to the harmonization of the overriding total context when viewed in terms of another deeper and more comprehensive sense. In a corresponding manner, as we have just seen, we can now say that it is only possible to judge the laws, rules, statistical hypotheses, and justificational theories utilized in a rational decision if we consider them from within the framework of that greater context to which they belong. Hence the rationality of these theories is in turn also something relative to a given situation, and thus only to be measured in terms of the degree to which it can be organized and subsumed within this comprehensive context.

I would like therefore to conclude this rather brief treatment of the 394 contemporary problems surrounding rational decision theory with the following two theses:

1. Rational decisions are historical in that they are always connected to historical conditions. They are determined by a situation; and therefore there exist no "intrinsic" rational contents which might take on the form of particular goals and assumptions. It is always the case that

goals and assumptions are only to be judged in terms of a given context, and accordingly they can never make a claim to being universally valid for all times, as is falsely asserted today by the advocates of a kind of new "enlightenment."

2. On the other hand, rational decisions are nonhistorical insofar as they always aim at the same thing and always take the same form: that is, their aim is to attain to an optimal agreement between the rational decisions and a comprehensive systematic totality. Obviously this does not mean conformity to some kind of facticity, to that which is simply given or simply exists. For since we are concerned here with the goal of genuine agreement or adequation, not to be achieved by accident, coercion, or through the ruse of some mere semblance, we must accordingly understand this goal as leading in part to a reasonable conservation of that which exists as well as in part to a reasonable alteration of this. This in turn means that we must consider such an agreement in terms of the given conditions of the historical context and not *in abstracto* or, as it were, in terms of some kind of allegedly eternal "rational ideals."

This then is the direction our investigation must take if we wish to discover the foundations for a modern theory of rational planning.

15

THE SIGNIFICANCE OF GREEK MYTH IN
THE AGE OF SCIENCE AND TECHNOLOGY

The owl of Minerva, Hegel reminds us, only begins her flight with the gathering dusk.ᵃ Hence, if today the foundations of science are indeed "reflected" and "reflected upon" with such decisiveness—and in fact the philosophical world seems to speak of little else—then this serves to show us that a naive sense of certainty has been lost. Let us then once again call to mind those questions which arise in conjunction with this reflective consciousness and which have been posed here as well in the course of our study: In what might the truth of scientific assertions and theories consist? What meaning do verifications and falsifications have with respect to this? How do we actually decide between two mutually contradictory theories? What constitutes scientific progress? What are the criteria by which the scientific can be distinguished, in any sense at all, from the nonscientific? It is obvious that all these questions are indicative of a shattered state of scientific legitimation (*eines gebrochenen wissenschaft-lichen Legitimitätsbewußtseins*). To recognize the extent of this, we need only look at the heated conflict engendered a short time ago by Kuhn's thesis concerning the structure of the processes belonging to scientific development.[1] The discovery of antinomies in the very midst of mathe- 396 matics at the beginning of the century now appears as little more than a preliminary tremor in comparison with the major crisis in which scientific reason finds itself today, even if this is still largely hidden from the general consciousness. More obvious, in that it is more immediately apparent, is the crisis in technological reason, indeed, in the whole scientific-technological world, which was discussed in the previous chapter. Such, then, is the present situation.

15.1 The Question of Justification as It Pertains to Myth; the Interrelation of Myth, the Numinous, and Art

As peculiar as it might initially seem, the following considerations concerned with Greek myth (*dem griechischen Mythos*) are inextricably related to the situation described above. This is the case because the mythical mode of viewing things is an alternative to science, even though today this mode of thinking is generally held to be something which has already run its course and is now historically "finished" or "closed off." That we are in fact dealing here with an alternative mode of thought is shown

---

a. The reference here is to the famous line from the end of the Preface to Hegel's *Grund-linien der Philosophie des Rechts*.

by the fact that science developed out of the destruction of Greek myth (generally abbreviated in what follows simply as "myth"). Accordingly, in a time when we are no longer quite so certain of the path we have carved out in the intervening 2,500 years, it again becomes important to take a look at this alternative position. In the present context we might state the meaning of this in the following manner: *The question of justification pertaining to science as that mode of thought which dominates almost everything today, a question which has recently become such a burning issue, cannot be treated without also addressing the question of justification as it pertains to myth.* In this sense, then, my comments on Greek myth are to be understood as contributing to the current discussion; for indeed, as Ernst Cassirer recognized and pointed out in full clarity earlier, such an object for study has to be considered an integral part of any theory of the sciences.[2] How are myth and science distinguished from one another? How can we choose between them? Or is it perhaps the case that the two merge at some point and are constantly flowing into one another? What gives us the right to assert the scientific point of view over the mythical? These are the questions which have then arisen unbidden and unforeseen within the framework of our contemporary situation; we must accordingly pursue them calmly and soberly, and without the numerous prejudices normally associated with this issue.

By proceeding in this manner we have in no way lost sight of the question of justification as it pertains to the object of art and the numinous realm—questions which were raised in the first chapter of this book. But the first chapter was meant as a *historical* introduction to the general problematic of what was to follow; and accordingly there we referred only to the *traditional* extrascientific modes of considering the world, namely, to religion and art. However, both of these traditional modes find their historical root in myth and in the fact that it was myth which dissolved and disintegrated into religion and art, disappearing in the process as a consolidated whole. At the same time, however, this disintegration only occurred under the pressure of the rising tide of science in the waning years of antiquity. It was only when the *logos* of Greek philosophy began to dislodge the mythical from the world that religion sought a relation to absolute *transcendence* and art was transformed into the *mere image* or *semblance* of the beautiful (wurde Kunst zum schönen *Schein*).[3] The mythical continued to live only in the numinous element of religion, where it became an object of increasing difficulty and embarrassment, and in the mythological content of art, where it was stripped of its essential efficacy and meaning for reality. In both of these areas it only survived in the fractured form induced by the "*logos*" and science.[4] However, in the modern world, where we are able to pose the question of justification for science in a much more radical form—as for instance Kant was able to do in seeking to give science a transcendental foundation—we are also able to confront science with this radical alternative and to compare and contrast the two—that is, to confront science with this radical alternative as it once existed in a form yet uninfluenced by science, as that consolidated whole in which religion and art were welded together. In other words, we are able to confront science with Greek myth as that utterly different and peculiar form of *immanent* world and reality experience which at the same time served as the historical point of origin for science. 399

What, then, constitutes this mythical form of experiencing the world and reality? How is it distinguished from that of science?

In order to find the essential clue which might serve to guide us in answering these questions, we must once again call to mind some of the results of chapters 4, 8, and 13. Here we can summarize as follows: The categories developed in chapter 4 for the natural sciences find a partial analogue (as chapter 13 showed) in the historical sciences. In a general sense that can be extended to cover both of these uses, the contents of these categories were called a priori precepts or principles. In this way we came to see (already in chapters 8 and 9) that the content of these precepts or principles did indeed for the most part admit of historical transformation, while a certain group of these precepts *defines* the scientific mode of thought (though it is rather obvious that this mode of thought as well could only be made understandable in a historical manner). To this latter group there belong, among other things, a few assertions of a highly general and universal nature (as we will show in what follows) concerning such things as causality, quality, substance, quantity, and time. But whereas the names designating these notions do indeed coincide with some of Kant's so-called categories of understanding and forms of intuition, here they have quite a different sense than that given them by Kant. In the first place, they are not, as in Kant, conditions of the possibility of any experience whatsoever, but rather only conditions of *scientific* experience. Second, for this reason what is to be expressed by them in the most universal way is limited to that which is *fundamental* to the scientific mode of considering things, which has been treated in the previous chapters. Thus we will only speak of certain formal aspects of the scientific conception of causality—namely, those which abstract from all particularities of content, like the concepts of determinism and indeterminism (cf. chapter 2). The same can be said for the scientific conception of time; hence problems of the type dealt with in chapter 10 will not be considered, etc. Kant, on the other hand, did in fact connect his categories and forms of intuition with the highly specific contents relating to Newtonian physics. 400

With this in mind, we can now pose the previous question more exactly in the following form: What, for instance, constitutes the *mythical* concept of causality, of quality, substance, quantity, as well as that of time? And what distinguishes these concepts from their *scientific* counterparts?

I admit that in speaking here about the relation between *the* mythical and *the* scientific conceptions, I am simplifying to a rather high degree. But nevertheless I see no real danger in doing this, since I limit myself here to the discussion of only a few *essential characteristics,* characteristics which on the one hand can be inferred from myth as a closed historical form that has run its course and on the other from science insofar as the latter admits of a historical overview of what has gone on up to the present—and the previous chapters have as well only been concerned with science in these terms. Here I have intentionally avoided the most recent developments in science, those which are presently still deeply involved in their formative stages, as is the case especially in the fields of microphysics and cosmology, even though these have in part led to results which bear a rather amazing similarity to mythical conceptions.

15.2 The Conditions of Mythical Experience

Let us then begin with the concept of *causality.* Mythically viewed, causality is a matter of divine agency, regardless of whether we are dealing with movement in place ($\kappa\alpha\tau\grave{\alpha}$ $\tau\acute{o}\pi o\nu$) or with qualitative transformation and metamorphosis ($\alpha\lambda\lambda o\acute{\iota}\omega\sigma\iota\varsigma$, $\mu\varepsilon\tau\alpha\beta o\lambda\acute{\eta}$). The casting of a spear, the coming of a storm or a wind, the movement of the clouds, the stars, and the sea—in all of these are revealed the powers of the gods. These powers are similarly active in the changing seasons, the outbreak of a plague, in the acquisition of knowledge, in inspiration, wisdom, self-control, delusion and infatuation, and suffering.[5] But this efficacy, variously directed at both locomotion and metamorphosis, exhibits certain typifiable characteristics. A god is not responsible for just any arbitrary activity, but rather for that activity which corresponds to his essential being: Helios causes the movement of the sun; Athena guides Achilles' spear in order to fulfill the historical destiny of the Achaeans whereas it is also her proximity or presence that acounts for practical prudence and clever council, just as that of Apollo is the source of prophetic vision and musical rapture; it is Aphrodite who enables man to fall in love, while Hermes is responsible for jesting and practical jokes, etc.[6]

Mythical causality can thus only be understood in the form of divine agency and efficacy when it is seen in relation to *the essential characteristics of the gods.* Such essential characteristics are *mythical qualities.* As we can then already see, even from the limited number of examples presented thus far, these qualities are primordial forms or structures (*Urgestalten*) and at the same time structural totalities (*gestalthafte Ganzheiten*) in that they are, to borrow a phrase from Walter F. Otto, a "multifarious being" (*ein "mannigfaltiges Sein"*).[7] They are elementary powers that constitute human reality; and their causal efficacy is conceived as an expression of their essence.

This becomes particularly clear in Hesiod. When Chaos gives birth to the two shadow-regions of Night and Erebos,[b] we obviously have something which springs from an essential relation between such qualities of darkness. When Night in turn gives birth to Day, it is again a quality, even if that of a polar opposite, that generates this causal sequence. A qualitative relation also exists between the Titans and the Olympian gods on the one hand and their primordial progenitors, Heaven (*Ouranos*) and Earth (*Gaia*), on the other, since both the Titans and the Olympian gods belong to the Earth as well as to Heaven. For further evidence we need only call to mind Prometheus, who brought fire down from the heavens; Mnemosyne (Memory), the guardian of divine wisdom; Themis, the protectress of divine order and justice; and her children, who again embody the qualities of righteousness and peace. We could go on to give a multiplicity of such examples and to show that a certain system is present throughout; but in the present circumstances we cannot take the time to elaborate on this in greater detail.

Nevertheless we can firmly assert the following: Mythical qualities are individual forms or structures which serve to shape the reality of man; and each of these individual structures presents us with a type for that efficacy which lies in its essence.[8] Thus, whereas in the mythical context causality is referred back to a quality, we find that in the scientific view things are for the most part reversed and qualities are derived from causal laws. Quality and causality admittedly mean something entirely different in science and myth; but nevertheless for the Greek imbued with mythical consciousness, the gods, as handed down to him by Homer and Hesiod in the form of primordial structures and primordial qualities, functioned as the alphabet which helped him, to use a Kantian phrase, to spell out his individual experiences just as certain universal and fundamental structures of causality and quality, taken here in a scientific sense, serve as a corresponding alphabet for modern man. The Greeks proceeded on the basis of a conception of the gods as primordial structures and primordial qualities; they saw them at work everywhere and experienced the world in terms of the spheres allotted to each of them, that is, within the framework of the types they personified, their order, and their causal relations and efficacies. Herodotus still sensed this when he wrote of Homer and Hesiod that they "were the first to compose genealogies and give the gods their epithets, to allot them their several offices and occupations, and describe their forms."[9] For the Greeks, then, these gods were the conditions of the possibility of mythical experience.[10]

403

404

b. Erebos (Erebus) is an intermediate region of the netherworld lying between Tartaros, the abyss of utter darkness, and the world of men. Hübner speaks of it here as an intermediate form of darkness, hence a kind of twilight realm.

One of the inextirpable errors of mankind is certainly to be found in the repeated assertion that the human experiential world is always necessarily one and the same. Thus, for instance, we hear it claimed that in previous ages the gods were supposedly invented a posteriori in order to explain universally recognized phenomena or to work out nice neat little stories about the gods, whereas later on everything came to be understood scientifically. But from a strictly scientific point of view it is precisely the opposite which turns out to be true; for according to this point of view there is absolutely no kind of noninterpreted experiential world, nor can there ever be such a world. The Greeks perceived the world *in the light* of their gods; and only by recognizing, to use Herodotus's words, their epithets, offices, occupations, and forms, did they initially come to *artic-*
405 *ulate* and order their world.[11] Everything else was then merely a consequence of this. We too understand everything in the light of a conception of causality and quality, albeit one involving completely different structures; and for us as well everything else, the individual and the particular, is merely a consequence of this. *The gods are the a priori of the mythical Greeks;* they make mythical experience possible. To this extent they are just as objective for them as are generally causal laws in the sciences and qualities which—in contradistinction to myth—are determined on the basis of these laws.[12]

Now, as we have already mentioned, all these mythical qualities are somehow personified, if they are not in fact persons. Hence they are individuals in space and time; and to this extent they are also *substances*. When Night brings with it sleep, death, dreams, etc., something of Night itself is present in these qualities, a darkness and nocturnalness. The same holds when Heaven and Earth unite to beget the Titans and the Olympian gods, for in these the heavenly and the earthly are united. Thus the substance of Night is in sleep and the dream, just as the substance of Heaven and Earth is to be found in the Titans and the Olympian gods—in the Sun, Fire, the order of Justice, and Custom, etc. These various parts and elements of Night, Heaven, and Earth, found in their offspring, differ no more from the *whole* of Night, Heaven, and Earth than does the redness of a surface from the redness of a piece of that surface: which is to say, mythically viewed there is no difference between a whole and its parts. This supplies us with a characteristic trait of *mythical quantity*.
406 This conception of mythical quantity—namely, that the whole is in every part, while at the same time both whole and part are each personified substances—makes it possible to understand how it is that a god can be in many places at the same time. Wherever prophetic wisdom, measure,
407 and order hold sway, there we find the Apollonian substance and, owing to the identity of whole and part, Apollo himself. Wherever beauty and grace cast their spell over the hearts of men, Aphrodite is present in person. And with respect to all of this, there are certain especially sig-

nificant moments, when, in a *Kairos*,ᶜ a person feels himself to be in the presence of the divinity. He feels this nearness when, as if transfixed by a ray of light, an enlivening force takes hold of him, when he himself is called a θεῖος. Here he senses the divine substance as it flows into him[13] just as we also find that the etymological origin of the word "influence" (*Einfluß*) embodies a meaning which comes down to us from the representational world of myth.[14]

There is another point of decisive significance here. Mythical quality 408
does not admit of a division into an ideal (*ideelle*)ᵈ and a material realm. Things which in our eyes belong to the ideal realm, like order, wisdom, measure, justice, delusion, love, etc., are viewed here simultaneously as *personified substances,* and hence as something material. Correspondingly, that which is material, like earth, heaven, the sea, or the sun, is also viewed as personified substance, and hence as something ideal. For this reason, in myth something spiritual can always be materialized or embodied and present itself in an individual shape or form, just as conversely something material might at any moment take on the characteristics of a person. Mythically seen, then, everything has a *holistic structure (ganzheitliche Gestalt)*. Hence, whether we are dealing with a primordial structure or a primordial quality, or with the causality which springs from the individual essences associated with these, whether it is a question of the relation of a whole to its parts, or of the ideal and the material—in all of this the mythical mode of thought, as a holistic mode, is *synthetic*. The *analytical* procedure as introduced into the world by science—that is, the forcible breaking down of the world (*Zerschlagen der Welt*) into abstract substances, atoms, and elements which move according to universal laws—is something that was completely foreign to the Greeks of this mythical age. I do not mean to imply by this that the difference between the ideal and the material was completely unknown to these Greeks; but what I do mean to point out is that for them the line separating these two spheres did not lie where we draw it today. And this we can assert because, as we have seen, they had a completely different conception of causality, quality, quantity, and substance.

No less astounding is the difference between their perception (*Anschauung*)ᵉ of time and ours. In order to make this clearer, I must once

c. The Greek word *kairos* (καιρός), when it refers to time, means something like "at the critical or opportune moment."

d. The word "ideal" (*ideell* and *Ideelles*) must be read here as the unadulterated adjectival form of idea—that is, as referring to the *invisible* world of objects contemplated by thought. German distinguishes, though somewhat tenuously, between the *Ideelle,* ideal in the above sense, and the *Ideale,* which carries a normative and paradigmatic value, e.g. the ideal person.

e. Hübner uses the term *Anschauung* in conjunction with time to bring out the relation to Kant's pure forms of *intuition,* space and time. However, as it is more normal to speak of the *perception* of time than of the *intuition* of time in ordinary English, I have generally again turn to the mythical conceptions of quality and causality. As stated earlier, mythical qualities, as divine primordial forms or structures, have their own specific kinds of efficacy. But both these structures and these

409 efficacies were *defined,* as it were, through particular tales or stories which were told about them, stories which in conjunction with Grønbech I will call *archai.*[15] An arche is a sacred event; it is the story of a god. The nature of each of the gods, that is, what he properly *is,* can only be gathered from his story. Here we find recorded his lineage, the account of his birth, and a description of his deeds. Some of these archai deal with natural occurrences, while others are more historical in nature. Accordingly, it seems appropriate to distinguish between natural and historical archai. For example, we find natural archai in Hesiod when he describes the genesis of the world from Chaos, Earth, and Eros; and again in the departure and return of Persephone, which parallels the change of seasons. To the historical archai belong stories like Apollo's slaying of the serpent Python, the battle of the Titans, Hermes' theft of the sacred cattle, Athena's bestowal of the olive tree, and the saga of Erechtheus.

Now, when Hesiod recounted the story of the genesis of the world, he was not thinking about events that took place *in* time; indeed, as far as I can tell, the word "time" in its characteristic form does not appear once in the whole Theogony. Here we find nothing of that separation between time and what it contains so familiar to us today. The natural archai— Chaos and the story of what it gave rise to, Earth and the tale of what she brings forth, the succession of Night and Day, Heaven and Earth, Mountains and Sea, the movement of the Sun, etc.—all of these are

410 primordial temporal elements (*zeitliche Urelemente*); they are not, as is the case in science, based on or related to some kind of abstract points or some segment of an invented time-continuum. Each of these archai is an *individual* story with a beginning and an end. Here the flow of time belonging to the world is primarily like the turning of pages, each of which is new, in the book of these cosmic tales, until we finally reach the point where they begin to repeat themselves cyclically over and over again. Each of these individual stories, as an arche and primordial mythical structure and quality, has its own internal form of succession; and each contains *within itself,* as an intrinsic element, the relation to the story

---

translated *Anschauung* as perception. But there are places, especially in the notes, where the reference to Kant is either directly stated or necessary for understanding the sense of a passage; in these places I have generally used *intuition* as the appropriate translation. Further, it is also important to realize that Hübner distinguishes between the *categories* of substance, quantity, quality, etc., on the one hand, and the *forms of possible intuition,* space and time, on the other. The use of *Anschauung* is meant to bring to mind this distinction as well; hence the reader is advised to remember that whenever the phrase "perception of time" occurs, we are dealing with something which, in Hübner's opinion, cannot simply be lumped together with other categorial structures.

which follows immediately upon it, while at the same time each and every one of these stories is absolute in that it cannot be reduced or referred back to anything else. It is these archai, and only these archai, that constitute time in the mythical sense. And it is for this reason that we are able to call them not only time-elements, but *time-forms* or *time-structures* *(Zeitgestalten)* as well.[16] Hence, what we separate in a *universal manner* from our scientific perspective, without distinguishing among various and sundry theories concerning *particular* causal laws, space-time structures, etc.—namely, the individual events, time conceived as a continuum of points, and the universal causal laws by means of which these events are sequentially ordered within time—are here fused together for the Greeks into an indissoluble whole, that is, into the wholeness and unity of an arche.[17]

411

We can thus say in the first place, then, that when we consider matters topologically, mythical time has an *absolute beginning,* just as we see in Hesiod. For, since the arche of Chaos is in fact not *in* time but rather constitutes time, it would be senseless to ask about a time before the arche of Chaos, just as it would be senseless to ask what lies beyond the edge of Einstein's curved universe. Second, we can state that, again when topologically viewed, mythical time is *cyclical,* insofar as at least a part of the natural archai, if not all, incorporate an *identical* recurrence. Among those archai where there can be no question about this, we can place the stories concerned with the birth of Day from Night, the circular course of Helios and the stars, and the eternal rhythm of the departure and return of Persephone which constitutes the mythical interpretation of the succession of the seasons. Here we are always dealing with the recurrence of the same event; the *very same* divine and sacred story is repeated over and over again. It is in this respect that the Greeks also speak of a notion of sacred time, ζάθεος χρόνος.

However, running through the whole of Greek myth like an unerring and unbroken line is the distinction between the sacred and the eternal on the one hand, the world of the gods and their archai, and the world of the mortals, the βροτοί, on the other. The Greeks correspondingly distinguish between ζάθεος χρόνος and human time, the latter simply being called χρόνος (a distinction which Fränkel pays particular attention to in his writings).[18] Contrary to the world of the immortals, the world of the mortals, the profane world, is characterized precisely by the fact that within it nothing recurs; everything changes and is transformed, or simply disappears without a trace. In this world it becomes necessary to hold on to the past and to calculate the future; and for this reason temporal serial enumeration, the progressive differentiation of discrete hours, days, and years, is of fundamental significance. In the profane or human world the ordering of events *in* time is unavoidable if these events are to be identifiable in any sense at all. Hence a constitutive factor of the battle of

412

Marathon is that it took place in 490 B.C. (or the Greek equivalent of this time designation). For the eternal return of the same as such, however, the number of times it recurs is not a constitutive factor. Thus, for example, the reappearance of Persephone requires no dating for its identification.

Therefore *profane* time flows from the past into the future, just as we are accustomed to think of it today. Past, present, and future are strictly separated. The past is irrevocably gone; the future is unknown. *Sacred* time, on the other hand, folds back into itself over and over again. That which is already past always recurs cyclically in the future; and the cycle itself, that is, the structure of the arche, is thus the eternal present. The mythical Greeks lived in a multidimensional reality that included the dimension of the sacred as well as that of the profane. In the dimension of the sacred, the archai shone forth for them as eternal primordial images, as *archetypes;* and they used these guiding stars to orient themselves within the profane realm. This they did first of all by utilizing the internal metrics of the archai, their rhythm, and second by serially enumerating their repetitions for the purpose of identifying that which is mortal, that which never recurs. To this extent, then, profane time is merely secondary and derived from sacred time in the ancient Greek conception: the all-encompassing cosmos, in any case "sacred nature," is not subjected to the structures of profane time. But even in the dimension of profane time the holistic mode of mythical thought is readily apparent, since the bifurcation of reality into two dimensions is not to be understood here in any way as a separation, but rather as just the opposite. These Greeks derived both temporal order and direction, as well as the temporal standard of measure for their mortal world, from the immediate *perception (Anschauung)* of the sacred, of the primordial events, the archai, that is, from the perception of the eternal return of that which is divinely the same, as manifest in the course of the heavenly bodies and the rhythm of the seasons.[19]

Here again, then, we find that the perception (*Anschauung*) of time is in fact the condition of the possibility of experience. But it is a perception of time which is wholly and completely different from that derived from the scientific point of view. From the mythical perspective this perception has to do with eternal time-structures and is a perception of an indissoluble totality encompassing both sacred and profane time. I should like to call this totality *mythical time*. The perception of time belonging to science on the other hand has been developed exclusively on the basis of *profane time*, even though with respect to certain particulars it is no longer identical with profane time in any sense at all. And for a rather lengthy period profane time was the absolute standard for the scientific perception of time. Everything was ordered and arranged within *it;* and whatever could not be so ordered, like the absolute time-structures of the archai, was declared nonexistent. One no longer saw the same, that which repeats itself identically, in the rhythm of the days and seasons. It was no longer *the* spring whose return was joyously celebrated. Rather, every unit of time which clicked off was viewed as something new, something unique that was never to recur again. With this, nature, as something sacred, disappeared and was transformed instead into something mortal.

Thus far we have only spoken of the natural archai in conjunction with mythical time. However, the historical archai also play an important role here. These archai are also time-structures to the extent that they present the particular and individual course of an event which had a first occurrence (τὰ πρῶτα) and then constantly recurred in an identical manner as precisely *this* individual event. Thus we find an historical arche functioning as a time-structure whenever some god bestows or establishes something *for the first time:* when Athena bestows the olive tree and establishes the art of weaving; when Apollo establishes the order of the state and bestows that of music; when Hermes establishes business, barter, and trade. Since all of these are both stories and sequences of events, both of which belong to the mythical quality and substance of the divinity in question, these substances are also active whenever and wherever people plant olive trees or use the loom, make music or engage in business, etc. Whenever such things occur the old arche repeats itself; the same primordial occurrence is played out again, and the corresponding god is present and is called upon or entreated. Indeed, here as well the eternal return of the same is presented along with the arche itself: For it is the will of Athena to "show" the olive tree ('Αθηνᾶ 'εδείκνυ) so that its planting and use will be imitated; to "show" weaving, so that it will be practiced; etc. It is an essential characteristic of the historical arche that, as a story, it is part of the mythical substance of the deity and in this respect quite literally flows into the hearts of men, and again and again becomes effective within them.

The historical archai are related to the natural archai in much the same way that laws and rules governing the activities of men, when viewed from our present-day perspective, are related to the laws of nature and the universe. In the realm of human activity too—at least in all the functions where the above-mentioned entreaty of a deity was demanded—we once again notice the absence of the three elements discussed earlier: we find neither individual discrete events, nor a separated time, nor anything like some kind of psychological causal law which might serve to arrange these events in their temporal sequence. What we do find here are once again the unique and individual archai, divine stories that constantly repeat themselves as self-contained structures and as such flow into the people as divine substance, where they continue to exercise an active power.[20]

416   Such archai were especially evident at the sacred festivals, where they were experienced with particular force. It would be naive for us to attempt to impose our own image of the world on early antiquity, for instance, to attempt to view the presentation of the Apollonian myth at Delphi, which had as its subject matter the slaying of Python, as a kind of theatrical production in which this event, which occurred in some primordial age, is supposedly merely being imitated and presented. Rather, what we have here is more appropriately to be seen as the *rite of a cult* in which a past event is retrieved and reaffirmed in the present where it actually takes place again. "It is no mere play," Cassirer writes in his work on *Mythical Thought,* "that the dancer in a mythical drama is enacting; the dancer *is* the god, he *becomes* the god. . . . What happens in these rites, as in most of the mystery cults, is no mere imitative portrayal of an event but is the event itself [in its immediate presence and power]; it is a δρώμενον, that is, a real and thoroughly effective action."[21] "Where we see mere 'representation' . . . myth sees real identity. The 'image' does not represent the 'thing'; it *is* the thing. . . . In all mythical action a true [tran]substantiation is effected at some moment; the subject of the action is transformed into a god or a demon whom it represents."[22] On the one hand, then, we find an event out of the distant past coming to expression

417   here in a presentation; and indeed the event draws part of its meaning, its gravity, reverence, and sacredness, precisely from this connection with the primordial past. On the other hand, however, this event, belonging so thoroughly and completely to the past, is expressly experienced as something immediately present that exhibits an awesome and powerful effect on the people viewing it. Wherever, as in the sphere of the sacred, the ideal and the real merge to form a unity—something which is possible in the mythical world because, as we have pointed out, mythical substance, quality, and causality do not admit of separation—we find that what to our modern gaze is merely a *representational* past is to the mythical consciousness an immediate present as well. Moreover, the future also becomes the present in such cases, something which is made possible by the knowledge that the eternal return of the same will be carried forward from sacred festival to sacred festival. Thus here we once again find that temporal distinctions disappear for the Greeks.[23]

    Sacred and profane time are not coherent: Archai and mortal events belong to different dimensions of reality, even if these dimensions are reciprocally tied together in an unbreakable relation. Further, there is in myth no unified topology of time by means of which its direction and order might be univocally defined, just as there is also no unified metric

418   under which everything might be subsumed. If we wish to use a modern expression drawn from mathematics and physics, we could then say, for example, that historical archai present topological as well as metrical singularities in profane time: topological singularities, because the order

and direction of time can be reversed in them so that they, as something past, are capable of effecting the present without the use of intermediary temporal elements; metrical singularities, because no particular duration is a constitutive factor necessary for their identification. (Thus, for instance, the duration of Apollo's battle with Python is obviously of absolutely no significance; and if any kind of temporal designations (days, hours, etc.) were ever mentioned with respect to this, they were not meant to be taken literally.)[24]

From what we have said thus far, we can already see that historical 419 archai, like natural archai, are conditions of the possibility of mythical experience. The natural archai constitute the mythical time which the Greeks used to orient themselves both within nature and in their day to day lives. The historical archai, on the other hand, those structures which enabled them to make the primordial past present in all its sacredness, lent them wisdom, served to guide them in their councils and actions, defined their customs and laws, filled them with power, granted them good fortune, and gave them an eternal sense of meaning and significance.[25] All of these archai, historical as well as natural, standing as visibly present structures, made it possible for these Greeks to break out of their profane world and enter into the perception of the sacred (*Anschauung des Heiligen*).

15.3 The Destruction of Myth through the Ascendance of Science
All of what we have said above can be seen in and inferred from practically every line of Hesiod, Homer, Pindar, or any other text that has come down to us from antiquity. It is also manifest in Greek art and in the sacred shrines and rites of that culture. But perhaps our clearest experience of all this comes from the study of those people who expended 420 such a massive amount of energy in the attempt to destroy the mythical mode of thought. Here I have in mind, above all, those ancient Greek scholars who were active as logographers, mythographers, and genealogists at the time of the rise of philosophy and science—notably, Hekataios, Pherekydes, Hellanikos, Xenophanes, Ephoros and many others. These men all contributed in various ways to the destruction of Greek myth; and probably most deadly of all was their attempt to arrange the tales and stories of the archai, as well as the personages and gods who appeared in them, in terms of a chronological system of profane time. The principal means to this end were the genealogies they devised. It is for this reason that their mythographic and logographic works also generally fall under the heading of γενεαλογίαι.

At first the genealogies of the world of saga and myth were merely isolated and fragmentary works which did not employ designations belonging to the temporal order. Later these were transformed into comprehensive family trees of the gods and heroes. And in the end they gradually began to incorporate exact dating. Today it is with utter amazement that we look back upon something which seems to have been so completely out of the ordinary, as well as at the primitive means by which this was initially accomplished. For example, at first the author simply began with his own time as a point of origin (ἐξ ἐμέ); later the Olympiads were brought into use. Even then it was quite some time before they began, in addition, to bridge the immense gap in time which existed between the world of saga and myth and their own times by filling it in with genealogies. (In this respect Hellanikos may have been particularly instrumental. In his ἱεραί, he used the succession of the priestesses of Hera as a basis for establishing a continuous serial enumeration of historical events.)

421   The strength of the resistance the genealogists encountered can be sensed when you observe the passion, zeal, and effort with which they directed polemical attacks against their fellow Greeks in their attempt to bring them around to something which must obviously have seemed completely foreign to their way of thinking—namely, to view all events as somehow stretched out in the linear series indicative of profane time, to arrange these events, lock them into place, and date them within such a framework. In this way, not only did profane time become the sole condition of experience, with its unitary order, direction, and metrics, but there was now correspondingly only *one* reality, profane reality. At least in part, the genealogists were seeking to preserve the content of the myths by arranging them within this new unitary time system. But as it should now have become quite apparent, this was a hopeless task which finally had to end with the sacrifice of the entire body of myth and with its being explained away as mere fairy tales.

Here we find one of the first great examples of that notion which would later come to such preeminence: σώζειν τὰ φαινόμενα, save the phenomena. Here, as always, this was carried out by means of the introduction of a new concept of experience and reality over and against which the asserted facts had to be measured and assessed. The fact that the archai did not admit of sequential arrangement, of being locked into place and dated, obviously did not disturb the mythically disposed Greeks in the least; from this we can infer that the truth and reality of these archai were in no way dependent upon such forms of arrangement for them. Every attempt at preservation and salvation of the kind described above presupposes that the status of something which has in fact become dubious can be bolstered and secured—in the present case, something which could only be done by means of a forced interpolation of genealogical causal chains. Hence the fact that this idea of salvation never even occurred once to the mind of mythical man can only mean that the truth of his

422   archai seemed to be immediately *present* to him: that the past was still *there,* still *existed,* for him like something eternal, something which could

be directly and immediately seen in nature, in the heavens, in his own actions, and especially in his *cult*ural festivals. How could all of this possibly stand in need of the explanations of the genealogists—especially since for mythical man it was precisely these structures which served in an inverse manner as the point of origin and the means for all explanation, in that they constituted the very condition of possible experience? Grøn- bech is therefore completely correct when he writes that we have to "revolutionize" our conception of time if we want to understand the mythical Greeks' conception of time. "We involuntarily think of time as a stream," he writes, "which flows out of an unknown past and incessantly runs up against an equally unknowable future."[26] But for the Greeks, "time was not a place for occurrences; rather it was the occurrences themselves."[27] "They see something we are not able to see, and thus their thoughts move in a completely different dimension, so that no common denominator can be found. In our eyes the Greeks lived on two planes. The time of the festival is not contained in the stream of time; rather it lies outside of this; or more properly stated, it lies above the everyday world, like a high plateau from which rivers flow down into the lowlands of the moment (*des Augenblicks*). It is out of this arche that time unfolds itself; here, in this sacred place . . . is created that which will be made into the progressive works of everyday life."[28] Finally Grønbech writes: "In considering the spiritual life of the Greeks, we must not only recon- sider our own concepts, but also realign our way of thinking about ex- perience."[29]

423

It would take us too far afield to supplement the previous analysis with a corresponding treatment of the concept of space involved in Greek myth. As it is, one can only suppose that this would involve no less of a deviation from our conception of space than we found to be the case with respect to the difference between the mythical conception of time and our own conception of time, as well as that between the mythical conceptions of the other categories discussed here and their modern counterparts. For this reason, I would then like to conclude by turning once again to the consideration of the relation which exists between science and myth.

15.4 The Relation of Science and Myth

Here we must then ask, how it is possible to choose between the a priori structures and status of myth and science? How can we decide between the mythical conceptions of causality, quality, substance, time, etc., and the corresponding scientific conceptions?

It is precisely the scientific mode of considering things, that is, to the extent that this makes science itself an object of study, which compels us to realize that in both cases we are concerned with those structures which initially make experience possible and consequently with something which simply cannot be judged on the basis of experience. There is no

intrinsic or absolute reality which could serve here as a *tertium compar-*
ationis, since it is always the case that reality is already being viewed
either mythically or scientifically, and thus things are always being experienced either mythically or scientifically. Moreover, the same can be
said for reason. Both experience and reason, and *thereby* the criteria for
truth and reality, are then, among other things, always codetermined in
advance by particular causal and temporal conceptions. Accordingly,
nothing could be further from the truth than to subsume myth under the
irrational, as is so often done, and to set it in opposition to science, as
that which is supposed to be somehow representative of the rational.
Myth also has its rationality, a rationality which functions within the
framework of its own concepts of experience and reason, as we have
delineated these in terms of categories and forms of intuition (perception).
(The fact that in the case of myth this rationality is not afforded a quasi-
absolutized status, as has occurred with technology, is another matter.)
In a corresponding manner, myth also has its own particular type of
internal systematic harmonization by means of which all phenomena are
arranged within its total systematic context, a "logic" of its own "alphabet" and of its own fundamental forms or structures. If we might be
allowed to speak metaphorically, this is made at least partially apparent
to us when we are struck by the brilliant luminous clarity of the ancient
Greek world. But from all of this we are now led to the following conclusion: *Mythical and scientific experience, mythical and scientific rea-*
son, are incommensurable in a certain sense. And here "in a certain
sense" means that while we can indeed compare them, as we have in fact
just done, while we can understand them as alternatives, nevertheless we
have no kind of overreaching standard by which to *judge* them. Every
judgment is always something which is already conditioned by either the
mythical or scientific standpoint.

Can it then be that we are actually prevented from making any decision
in this matter? To this question there can be only one answer: The matter
has in fact already been decided for us for some thousands of years now.
But while this is undeniably true, it does not mean that we should make
light of the reasons underlying this awesome transformation, nor does it
mean that we should view everything merely from our own standpoint.
But as we have shown here, it is of no help to refer to *universalized*
concepts of experience, reason, truth, and reality in the attempt to understand this. Accordingly, we find that once again we must think of the
transition from myth to science as a *mutation,* in the sense attributed to
this concept in chapter 8, and hence as something which is system-historical (*systemgeschichtlich*). In doing so, we are obviously not permitted
to lose sight of the fact that this occurrence can only be conceived conditionally. For it would be just as impossible for the person imbued with
mythical consciousness to conceive of his divine world as an a priori for

the experience of the world in general, in the sense given this notion of the a priori in modern theory, as it would be for him to consciously think along lines which would permit him to impute the system-historical mode of thought to the principal historical figures of his time. So it is that we must unavoidably view myth to a certain extent today from a kind of *external perspective;* from its own internal perspective, it would inevitably paint quite a different picture of things. Thus a gap opens up here. But concerning this gap we can know at least one thing, namely, that it can never be filled in a way which would allow everything to be viewed as a continuous whole. Incommensurables can never be completely and adequately mediated.

Therefore an essential characteristic of this scientific mode of thought is to be found in the fact that, while on the one hand it will never be able to deny legitimacy to myth completely, on the other it can nevertheless make the historical destruction and transformation of myth rationally conceivable in its (science's) own terms—that is, as something which is conditioned in a system-historical manner. We cannot, and will not, simply turn back to the mythical mode of thought. It is something to which our entire realm of experience would be foreign and unintelligible, since the latter is determined in terms of completely different notions drawn from science. Nevertheless, we come to see that the burning question so heavily discussed today concerning the status and nature of truth in science forces us to a more serious reconsideration of the mythical, since the above question implies the further question of the truth of myth. This in turn leads to a more serious consideration of the numinous and art, as both 426 of these, as we mentioned at the beginning of this chapter, find their common root in the mythical. In any case, there is absolutely no *theoretically* cogent reason for assuming that mankind, even in the remote future, must relegate all mythical modes of considering the world as such to the realm of fairy tales, that is, when these are freed from the particular historical conditions of Greek myth. Nor is there any theoretical reason for supposing that if the prevailing opinion concerning this matter should change, mankind would have to forfeit rationality and, as it were, go insane.[30] However, we can say that no one today can predict whether, and in what manner, our horizons for viewing the world might actually be changed in the future so that the mythical could again become a living force and a new realm of experience. But this much we can assert with certainty: It is important to recognize this sheer *possibility* and to keep in mind that such a change could take place in that moment when the power of the one-sided technological-scientific world in which we live becomes less evident and imposing than it has been, while the questionableness of this world becomes more readily visible than has hitherto been the case.

NOTES

Chapter 1

1. V. I. [W. J.] Lenin, *Materialismus und Empiriokritizismus* (Berlin, 1958), p. 180.

2. Immanuel Kant, *Critique of Pure Reason,* B 236, tr. N. Kemp Smith, p. 220.

3. H. Reichenbach, *Wahrscheinlichkeitslehre* (Leiden, 1935), p. 420.

4. R. Otto, *Das Heilige: Ueber das Irrationale in der Idee des Göttlichen und sein Verhältnis zum Rationalen* (München, 1936). [Cf. *The Idea of the Holy: An Inquiry into the Nonrational Factor in the Idea of the Divine and Its Relation to the Rational,* tr. J. W. Harvey (London, 1926).]

Chapter 2

1. W. Heisenberg, "Ueber den anschaulichen Inhalt der quantentheoretischen Kinematik und Mechanik," *Zeitschrift für Physik,* vol. 43 (1927), p. 197.

2. Ibid.

3. Ibid.

4. W. Stegmüller, "Das Problem der Kausalität," in *Probleme der Wissenschaftstheorie, Festschrift für Victor Kraft* (Vienna, 1960), p. 183.

5. W. Heisenberg, *Physikalische Prinzipien der Quantentheorie* (Mannheim, 1958), p. 45.

6. C. F. von Weizsäcker, *Zum Weltbild der Physik* (Stuttgart, 1958), p. 85f.

7. David Bohm, in *Physical Review,* vol. 85 (1952), n. 2, p. 187. Here it is of no interest that this example, as well as that of the Copenhagen school, might also almost be called "historical." Both of these examples should be seen as case studies in which certain theoretical-epistemological fundamental orientations (*erkenntnistheoretische Grundeinstellungen*) should be demonstrated and tested. They should also indicate how these orientations are connected with particular philosophical theories.

8. D. Bohm, in *Physical Review,* vol. 85 (1952), no. 2, p. 166ff.; ibid., vol. 89 (1953), no. 2, p. 458ff.; *Progr. of Theoretical Phys* vol 9, no. 3 (1953), p. 273ff.; and J. P. Vigier, in *Physical Review,* vol. 96 (1954), no. 1, p. 208ff.

9. W. Heisenberg, "The Development of the Interpretation of the Quantum Theory," in *Niels Bohr and the Development of Physics,* ed. W. Pauli (London, 1955), p. 17f.

10. D. Bohm, *Causality and Chance in Modern Physics* (London, 1958), p. 170.

11. Heisenberg, "Development of the Interpretation of the Quantum Theory," p. 18.

12. With the aid of a partition of a large number of equal systems and a statistical count, one arrives at resultant differences of the following kind:

$$| \Psi(q,t_1)|^2 - | \Psi(q,t_0)|^2 .$$

This is an approximation of

$$\frac{\delta}{\delta t} | \Psi(q,t_0)|^2 .$$

With the aid of particular mathematical methods, we can derive a Ψ-function that satisfies the data arrived at through the statistical count for $|\Psi(q,t_0)|^2$ and for $(\delta/\delta t)|\Psi(q,t_0)|^2$, and *at the same time* presents a solution for the second Schrödinger equation.

This is then the "experimentally determined" Ψ-function. Accordingly, this determination indicates precisely that it is only possible as a rough approximation and that it is not susceptible to an exact measurement.

13. An argument has been made against this formulation in the name of empiricism which states that this formulation rests on a too narrow, and thus inappropriate, determination of the concept of empirical verifiability. If one followed the arguments in Carnap's treatise, "Theoretical Concepts of Science" ("Theoretische Begriffe der Wissenschaft," *Zeitschrift für Philosophische Forschung,* vol. 14 (1960), p. 209ff.), then a combination of universal and existential statements of a type like the causal principle would also prove to be "empirically significant," and thereby not without empirical content.

I cannot here go into the particulars of the work of Carnap cited above. In brief, however, let me say the following: In this work Carnap has expressly given up his earlier claim—which was still prominent in "Testability and Meaning" (*Philosophy of Science,* vol. 3/4 [1936/37])—that all theoretical predicates and propositions must be reducible, either completely or incompletely, to the immediately observable. In place of this, his "empirical significance criterion" now states in brief the following: A theoretical concept is significant if its application within a "particular hypothesis" results in a change in the prediction of an observable event. Thereby such a concept appears within a "theoretical language," which yields, by means of arbitrary "coordinating rules," a merely indirect connection to propositions concerning the directly observable, from which this language is strictly separated. (This is a conception which, in my opinion, rests upon a fundamentally correct analysis of scientific theory.)

Carnap's general change from a theoretical-perceptual beginning point to one that has only the practicability of science as its guiding idea is unmistakable. Hence I ask: What does this significance criterion still have to do with empiricism? We can no longer speak of a complete or even incomplete resolution of concepts into observable predicates; for us the only decisive point is that theoretical concepts achieve their codetermining role in the prediction within the context of observational statements, theoretical postulates, and coordinating rules. Such concepts thus arise from the spontaneity of thought, not from perception—and for this reason are not of an empirical nature.

Therefore we can see that Carnap's treatise does not constitute an improvement in the empirical standpoint over "Testability and Meaning" and the Vienna Circle, but rather that this treatise is the definite renunciation of this standpoint.

I wish to counter the above-mentioned objection that the causal principle proves to be empirically meaningful and practically verifiable because of Carnap's treatise "Theoretical Concepts of Science." I would argue as follows: This objection rests for its part on an inappropriate determination of the concept "empirical." For we can see from Carnap's article that there, in contrast to his earlier empirical orientation, he has recognized the decisive significance of the spontaneity of thought, which is stimulated, but not determined, by observances in the entire realm of science (not only in that of logic).

In this context it is instructive to consider what Carnap understood by saying that something is explained as "real" ("*wirklich*") in the realm of science, namely, merely that a theory is recognized insofar as its postulates together with its coordinating rules can be used to guide expectations or, more exactly, to derive the observational statements that express these expectations. Hence for him, the "real" was that which we use for the *practical* purpose of setting goals, but not that which theoretically rests on perceptions.

Cf. also W. Stegmüller, "Das Problem der Kausalität," in *Probleme der Wissenschaftstheorie, Festschrift für Victor Kraft* (Vienna, 1960), p. 87f.; and A. Pap, *Analytische Erkenntnistheorie* (Vienna, 1955), p. 138f.

Chapter 3

1. On this, see B. Riemann, *Ueber die Hypothesen, welche der Geometrie zugrunde liegen* (Göttingen, 1892), no. 13; H. Poincaré, *La science et l'hypothèse* (Paris, 1925); A. Einstein, *Geometrie und Erfahrung* (Berlin, 1921); H. Dingler, *Relativitätstheorie und Oekonomienprinzip* (Leipzig, 1922); H. Reichenbach, *Philosophie der Raum-Zeit-Lehre* (Berlin, 1928) [*The Philosophy of Space and Time*, tr. M. Reichenbach and J. Freund (New York, 1958)]; A. Grünbaum, *Philosophical Problems of Space and Time* (New York, 1963); 2d enlarged ed. (Dordrecht and Boston, 1973).

2. On this, see P. Duhem, *La theorie physique: Son objet, sa structure* (Paris, 1914) [*The Aim and Structure of Physical Theory*, tr. P. Wiener (Princeton, 1954)]; E. Cassirer, *Das Erkenntnisproblem in der Philosophie und Wissenschaft der neueren Zeit von Hegels Tod bis zur Gegenwart* (Stuttgart, 1957) [*The Problem of Knowledge: Philosophy, Science, and History since Hegel*, tr. W. H. Woglom and C. W. Hendel (New Haven, 1950)]; R. Carnap, "Theoretische Begriffe der Wissenschaft," in *Zeitschrift für philosophische Forschung* (1960).

3. We are only given n number of individual measurements, each of which deviates from the others: $1_1, 1_2, \ldots, 1_n$; however, the "true value" X, the very existence of which depends on a precept, is never given. Let

$$e_k = 1_k - X$$

be the deviation of the particular values from the assumed true value X. If we make the further assumption that the error might be either plus or minus with equal probability (in which case its algebraic sum approximates a vanishing point), then the true mean error for the particular value of 1_k can be calculated as

$$\mu = \sqrt{\frac{1}{n} \sum_{1}^{n} e_k^2} .$$

Finally, we then make the further presupposition that the arithmetic mean of the particular values—called the optimum value (*Bestwert*) L—comes closest to the true value. Correspondingly, by using $v_k = 1_k - L$ as the mean error with respect to L, we arrive at

$$\frac{1}{n} \sum_{1}^{n} v_k^2 .$$

Utilizing all these nonempirical precepts, we then arrive by simple operations at the equation

$$\mu = \sqrt{\frac{v}{n-1}},$$

where

$$v = \sum_{1}^{n} v_k^2.$$

And if we designate ΔL as the mean deviation of L from X, we finally arrive at

$$\Delta L = \frac{\mu}{\sqrt{n}}.$$

(For a more detailed analysis, see W. Westphal, *Physikalisches Praktikum,* 11th ed. [Braunschweig, 1963], p. 290f.)

4. If, for example, one chooses the Newtonian interpolation formula in order to determine more precisely the function coordinating a set of paired values x,y that are given through measurements, then one has already presupposed that this should be an entirely rational function (*eine ganze rationale Funktion*).

5. Einstein can be taken as an excellent representative of this point of view. Cf. A. Einstein, "Zur Methodik der theoretischen Physik," in *Mien Weltbild* (Berlin, 1960) [*The World as I See It,* tr. Alan Harris (New York, 1949)].

6. This conception also seems to represent the view of von Weizsäcker in his book *Zum Weltbild der Physik,* 7th ed. (Stuttgart, 1958). Concerning this, also see K. Hübner, "Beiträge zur Philosophie der Physik," in *Philosophische Rundschau,* Beiheft 4 (Tübingen, 1963).

7. Cf. T. S. Kuhn, *The Structure of Scientific Revolutions* (Chicago, 1962), p. 100ff. In the years since Kuhn's work first appeared, a rather comprehensive literature dealing with the topic of "the limited case" (*Grenzfall*) has also appeared, which I am not able to go into here. However, I do not have the impression that this work has produced anything decidedly new to date.

8. K. R. Popper, *Logik der Forschung,* 2d ed. (Tübingen, 1966) [*The Logic of Scientific Discovery,* 2d, rev. ed. (New York, 1968)].

9. This is a generalization of ideas already used by Poincaré, Reichenbach, and Einstein in treating the relation of geometry and experience. Cf. bibliography given in note 1 of this chapter.

Chapter 4

1. P. Duhem, *La théorie physique: Son objet, sa structure* (Paris, 1914). However, because the French edition is out of print, in the following I make reference to the English translation: *The Aim and Structure of Physical Theory,* tr. P. Wiener (Princeton, 1954).

2. Ibid., p. 268f.
3. Ibid., p. 134.
4. Ibid., p. 148.
5. Ibid., p. 166.
6. Ibid., p. 183f.

7. Ibid., p. 216f.

8. Ibid., p. 261f.

9. Ibid., p. 26.

10. Ibid., p. 335.

11. Ibid., p. 220f.

12. Cf. (among others), H. Blumenberg, *Die Kopernikanische Wende* (Frankfurt a. M., 1965); also *Die Genesis der Kopernikanischen Welt* (Frankfurt a. M., 1975).

13. It can be said that the Aristotelian philosophy contradicts the idea of unity in the sense denoted here, precisely in its fundamental principle. There we are concerned primarily with particular forms and qualities, which serve as final principles of explanation and which cannot be deduced from universal laws (and as is well known, we find very few laws such as these in Aristotle). In this we also see that this constantly growing accumulation of new qualities, which Aristotelianism was increasingly forced to accept, provoked no opposition.

14. K. R. Popper, *The Logic of Scientific Discovery*, 2d rev. ed. (New York, 1968). [*Logik der Forschung*, 2d rev. ed. (Tübingen, 1966).]

Chapter 5

1. "Nobis cum divina benignitas Tychonem Brahe observatorem diligentissimum concesserit, cujus ex observatis error hujus calculi Ptolemaici VIII minutorum in Marte arguitur; aequum est, ut grata mente hoc Dei beneficium et agnoscamus et excolamus. In id nempe elaboremus, ut genuinam formam motuum coelestium (his argumentis fallacium suppositionum deprehensarum suffulti) tandem indagemus. Quam viam in sequentibus ipse pro meo modulo aliis praeibo . . . , sola igitur haec octo minuta viam praeiverunt ad totam Astronomiam reformandam, suntque materia magnae parti hujus operis facta." Johannes Kepler, *Gesammelte Werke,* ed. W. v.-Dyck and M. Caspar (Munich, 1937ff.), vol. 3, "Astronomia Nova," chapter 19, p. 178.

2. Ptolemy and Copernicus set the upper limit for the margin of error at 10 minutes.

3. For a discussion of the philosophical background of Copernicus, see (among others) H. Blumenberg, *Die Kopernikanische Wende* (Frankfurt a. M., 1965) and *Die Genesis der Kopernikanischen Welt* (Frankfurt a. M., 1975).

4. Here one should recall briefly the most important arguments of Copernicus: The Sun does not move, but rather stands still because rest is more befitting its divine nature than the movement characteristic of lower orders; the rotation of the Earth about its axis is the result of its spherical shape and corresponds to its substantial form; thus viewed, this rotation is a natural movement for the Earth, for reason of which no centrifugal forces are apparent on it, as these appear only in forced motion; and most important, all things participate in the movement of the Earth owing to their "earthliness." One sees everywhere here how Aristotelian metaphysics has been either simply turned around or set in opposition to itself. Copernicus was especially fond of pointing to the fact that in comparison with Ptolemy his own theory was simpler. However, this is only true in a very restricted sense. As already indicated, Copernicus did not arrive at thirty-four epicycles, but actually required forty-eight. In his book *Galilée et la loi d'inertie* (Paris, 1939), Alexandre Koyre has amply depicted the history of the principle of inertia as this followed from the Copernican system.

5. The purely formal part of my exposition employs the work of R. Small, *An Account of the Astronomical Discoveries of Kepler* (a reprinting of the 1804 text [Madison, 1963]), and E. J. Dijksterhuis, *Die Mechanisierung des Weltbildes* (cf. note 1, chapter 4). The presentations in the "Astronomia Nova" are too involved

for the purposes of the present text; accordingly Kepler's ideas are only recounted here in their principal features.

6. The calculation of the parallax follows a method used by Copernicus.

7. Just as little as the Ptolemaic system was geocentric in a strict sense.

8. "Scopus meus hic est, ut Caelestem machinam dicam non esse instar divinj animalis, sed instar horologij (qui horologium credit esse animatum, is gloriam artificis tribuit operj), ut in quae pene omnis motuum varietas ab una simplicissima vi magnetica corporalj, utj in horologio motus omnes a simplicissimo pondere." Letter of Kepler to Herwart v. Hohenburg, Feb. 10, 1605; contained in Kepler, *Gesammelte Werke,* vol. 15, "Briefe 1604–1607," no. 325, p. 146.

9. "Quem cum viderem esse 100429, hic quasi e somno expergefactus, et novam lucem intuitus." Kepler, *Gesammelte Werke,* vol. 3, "Astronomia Nova," chapter 56, p. 346.

10. Kepler supposed that the Earth pulled all bodies along in the direction of its attractive force. He thought this force exerted itself in countless bands describing cones. The attracted object had its position at the tip of a cone as it touched the Earth. The resultant of these forces is vertical and directed downward. By means of this, Kepler sought to explain why a greater resistance is noticeable when a body is lifted up than when it is pushed sideways. But once again this obviously did not explain why the movement of a body in a direction opposite that of the Earth's rotation is possible without increased resistance.

11. K. R. Popper, *The Logic of Scientific Discovery,* 2d rev. ed. (New York, 1968), p. 33. [*Logik der Forschung,* 2d rev. ed. (Tübingen, 1966), p. 8.] (Page references to the second English edition are also accurate for the first English edition, which was published in 1959 [London: Hutchinson & Co.; New York: Basic Books].)

12. Ibid., p. 101f. [German, p. 67f.] It is nevertheless interesting to look more closely at the basic statements which play a falsifying role in Kepler. Here we are concerned initially with data which, moreover, have been partially constructed on the basis of theories and cannot be observed immediately (as, for example, the heliocentric longitudes). With the help of these data, he then determined other data, namely, the distances of stars, the positions of points in space, etc. All of this can just as well be expressed in basic statements that do not relate to observables. Kepler then rejected particular hypotheses by demonstrating that they could not be correlated with these data. Now Popper expressly demands that basic statements should be intersubjectively testable by means of "observation" (ibid., pp. 102–3 [German, p. 68]); but at the same time he adds to this that "observable" is an undefined fundamental concept, and it is sufficient to maintain that every basic statement must be about relative positions of physical bodies or equivalent to some basic statement of this "mechanistic" type (ibid., p. 103 [German, p. 68]). Therefore, even if Kepler's basic statements are basic statements in the sense of Popper, they clearly disclose how cautiously we should treat the word "observable."

13. Ibid., pp. 86–87. [German, p. 54.]

14. Ibid., p. 87, note 1. [German, p. 54, note 1.]

15. Ibid., p. 87, note 1. [German, p. 54, note 1.] Here it is worth noting that for Kepler an *experimentum crucis* between the Ptolemaic theory and his own was not possible owing to the minuteness of eccentricity in the planetary orbits.

16. Ibid., p. 268. [German, p. 213.] The apodicticity of this rule—overlooking the fact that the example of Kepler shows how unproductive such a rule can be for the process of science—also explicitly contradicts Popper's expressed declaration that basic statements, or, more exactly, hypotheses of a lower general

level, for their part constantly allow further testing and thus could never make claim to finality. Does it not follow from this that along with these basic statements or hypotheses the falsification decision concerning a theory could again come up as a problem?

17. Ibid., pp. 42, 81–84. [German, pp. 16, 48–51.]

18. Ibid., p. 131. [German, p. 93.]

19. Ibid., p. 130. [German, p. 92.]

20. Ibid., p. 130. [German, p. 92.]

21. Popper sees that nothing can force us to stop and hold fast to certain basic statements and that accordingly every basic statement can be examined again. Since all we can do is temporarily stop at some place or other, it would be best if we stopped where the testing is "easy." He writes: "If some day it should no longer be possible for scientific observers to reach agreement about basic statements this would amount to a failure of language as a means of universal communication. It would amount to a new 'Babel of Tongues': scientific discovery would be reduced to absurdity. In this new Babel, the soaring edifice of science would soon lie in ruins" (ibid., p. 104 [German, p. 70]). The question is then obviously whether there is any objective criterion at all for the expression "easy to test" ("leichte Nachprüfung"). Precisely because of their dependence on theory, basic statements do not express absolute facts that could necessitate absolute judgments. As a rule, the recognition of a falsifying basic statement is a rather complicated and involved procedure, which is anything but unproblematic. Nevertheless, science will not therefore cease its activity (fall into ruins)—and justly so, since the call for uniformity among scientists is dogmatic.

22. Had Kepler been able to decide between his own and the other contemporary theories by means of Popper's grades of testability, he would have had to give preference to the circular hypothesis because it has fewer dimensions (by which, as discussed, Popper understands the number of parameters by which a theory might still be falsified). But this criterion for decision making would also have been completely useless to him (the other, namely the member-class relation, does not apply here), since all of the theories in question had indeed already been falsified for some time (including his own). In fact this is always the case. Theories are not on display like cars in a showroom which have not been run, so that we can weigh their merits first and then try one out; rather they are always already running, and from the first moment on they all manifest their relatively lesser or greater degrees of inadequacy.

23. I. Lakatos, "Falsification and the Methodology of Scientific Research Programmes," in *Criticism and the Growth of Knowledge,* ed. I. Lakatos and S. Musgrave (Cambridge, 1970), p. 118.

24. Cf. I. Lakatos, "History of Science and Its Rational Reconstructions," in *Boston Studies in the Philosophy of Science, 1971,* ed. R. C. Buck and R. S. Cohen (PSA., 1970), p. 22. There Lakatos states what should be done with theories that cannot be sanctioned according to his rules.

25. R. Carnap, *Induktive Logik und Wahrscheinlichkeit,* ed. and reworked by W. Stegmüller (Vienna, 1959), p. 84f. Hereafter abbreviated as *I.L.*

26. P. A. Schilpp, ed., *The Philosophy of Rudolf Carnap, The Library of Living Philosophers,* vol. 11 (London, 1963), p. 980. Hereafter abbreviated as *Phil. of C.* More precisely, there it is not a matter of Carnap's, but rather Burks's notion, with which Carnap expressly agrees.

27. At first glance this might be taken to mean that we are concerned here with the inference which lies at the basis of the law of the radii, in terms of the system of inductive logic, with a universal inference (*Allschluß*) (namely, that inference

drawn from the velocity of the planets as functions of their distance from the Sun in two cases to a universalization of this for all cases, and thus to the level of a universal law). But the difficulty is that according to Carnap's system of inductive logic in this case the equation would read $c^*(h,e) = 0$. (Cf., here, *I.L.*, p. 226.) Carnap sought to overcome the arguments raised against this (for would not this mean that no natural law whatsoever could be confirmed?) by interpreting the practical sense of such a law differently. When someone formulates a law, then, according to Carnap, he does not mean by this that it will be valid in an infinite number of possible cases (here, for all points on the orbital path), but rather only in a finite set which is suited to human limitations and practical goals (here, for practical goals, a sufficiently large number of points on the orbital path). In my opinion this means that universal inferences of the kind indicated must be interpreted as predictive inferences (*Voraussageschlüsse*), and so as inferences from one observed random sample to another (not observed). (Other possible forms of inductive inference, the direct inference, inference by analogy, and inverse inference, are not taken into consideration here.) However, for predictive inferences the following formula holds according to Carnap:

$$c^*(h,e) = c^*(h,e') = \frac{\prod_{i=1}^{r} \binom{s_i + s_i' + w_i - 1}{s_i'}}{\binom{s + s' + \mu - 1}{s'}}.$$

Cf., here, *I.L.*, p. 226.
28. *Phil. of C.*, p. 972ff., 986.
29. *I.L.*, p. 86.
30. Ibid., p. 86.
31. Ibid., p. 87.
32. Ibid., p. 80.
33. Ibid., p. 97.
34. *Phil. of C.*, p. 978.
35. Ibid., p. 990.
36. *I.L.*, p. 8.
37. Ibid., p. 10.
38. Ibid.
39. *Phil. of C.*, p. 973. Despite the separation of the system of inductive logic and methodology, Carnap has nevertheless given a detailed discussion of five rules according to which practical decisions are made on the basis of considerations belonging to inductive logic. He discarded four of the rules because of their extremely limited validity; but he indicated a fifth as acceptable. It runs: "From among the possible ways of acting (*Handlungsweisen*), you should choose that one for which the estimation of the resultant gain is maximal" (Cf. the section in *I.L.*, pp. 108–24).
40. At this point we might also recall Lakatos's peculiar conception that in hindsight we are at least able to determine whether a theory, in this case that of Kepler, has been progressive. At the most, however, all we can really say is that for Newton Kepler was progressive in that Newton gave a new meaning to Kepler's results. And besides, what good is a rule that we can only employ in hindsight?

41. This succinct saying, which has since been quoted many times, was used by this author for the first time in 1969 ("Was zeigt Keplers 'Astronomia Nova' der modernen Wissenschaftstheorie?" in *Philosophia Naturalis,* vol. 11); whereas Lakatos first used the saying in the same form in 1970 ("History of Science and Its Rational Reconstruction," *Boston Studies in the Philosophy of Science,* vol. 8). This proves that the saying was indeed conceived by both authors independently of one another.

Chapter 6

1. A. Einstein, B. Podolsky, and N. Rosen, "Can Quantum-mechanical Description of Physical Reality Be Considered Complete?" *Physical Review,* vol. 47 (1935), p. 777.

2. Ibid.

3. Ibid.

4. According to the uncertainty relation both quantities are not simultaneously measurable.

5. Only one of the two mutually exclusive quantities can be determined in accordance with the uncertainty relation.

6. N. Bohr, "Can Quantum-mechanical Description of Physical Reality Be Considered Complete?" *Physical Review,* vol. 48 (1935), p. 700.

7. Ibid., pp. 701–2.

8. Cf. K. M. Meyer-Abich, *Korrespondenz, Individualität und Komplementarität: Eine Studie zur Geistesgeschichte der Quantentheorie in den Beiträgen Niels Bohrs* (Wiesbaden, 1965). This work gives a presentation of the development of Bohr's fundamental philosophical concept and also contains an extensive bibliography.

9. The situation is described in similar terms in P. K. Feyerabend's excellent article, "Niels Bohr's Interpretation of Quantum Theory," *Current Issues in the Philosophy of Science,* ed. H. Feigl and G. Maxwell (New York: Holt, Rinehart & Winston, 1961), pp. 372–90, 398–400. There we find the following: "I would like to repeat . . . that Bohr's argument is not supposed to *prove* that quantum-mechanical states are indeterminate; it is only supposed to show under what conditions the indeterminacy of the quantum states can be made compatible with the EPR" (p. 384). In regard to the difference of opinion between Bohr and Einstein, I would also cite the following passages from this article: "Now a closer analysis of the argument," that of EPR, "will show . . . that it is conclusive only if it is assumed that dynamical states are *properties* of systems rather than *relations* between systems and measuring devices in action" (p. 381). Later on (p. 383) Feyerabend writes that Bohr was able to defend himself against Einstein by supposing "that states are *relations* between systems and measuring devices in action rather than properties of such systems." Feyerabend also points out that Einstein is not able to determine the quantities that he thinks exist in themselves and Feyerabend presumes that simply *stipulating* certain values in such cases would entail the violation of the principle of the conservation of energy. But Einstein was not directly concerned with either of these problems. His primary concern was to show the possibility of a completely different interpretation of quantum mechanics than the one in favor at that time and hence to stimulate new theoretical considerations, even if it might be difficult to specify their consequences for the moment.

10. P. A. Schilpp, ed. *Albert Einstein, Philosopher-Scientist* (Evanston, Ill.: Open Court, 1949), p. 669.

11. D. I. Blokhintsev, "Kritik der philosophischen Anschauungen der sog. 'Kopenhagener Schule' in der Physik," *Sowjetwissenschaft, Naturwissenschaftliche Abteilung,* vol. 6 (1953), Heft 4, p. 551.

12. Cf. K. Hübner, "Beiträge zur Philosophie der Physik," *Philosophische Rundschau,* Beiheft 4 (1963), pp. 74–78.

13. In *Die Naturwissenschaften,* vol. 23 (1935), pp. 808–12, 823–28, 844–49.

14. Schilpp, ed., *Albert Einstein,* pp. 669–73.

15. *Zeitschrift für Physik,* vol. 133 (1952), pp. 101–8.

16. H. Reichenbach, *Philosophische Grundlagen der Quantenmechanik* (Basel, 1949), p. 36.

17. Ibid., §7 & §8 and §25–§27.

18. J. von Neumann, *Mathematische Grundlagen der Quantenmechanik* (Berlin, 1932).

19. Let us suppose that

$$\Psi = \sum_k c_k \phi_k \, ,$$

where ϕ_k are the eigenfunctions of quantity A and α_k the eigenvalues of A. If N systems have the state Ψ (pure case), then it is predictable that in future measurements of quantity A we will get the value α_1, $N \mid c_1 \mid^2$ times the value α_2, $N \mid c_2 \mid^2$ times, etc. Now take χ_l as eigenfunctions of the quantity B, β_l as its eigenvalues, and assume further that $[AB - BA] \neq 0$. Then we have

$$\Psi = \sum_k c_k \phi_k = \sum_{kl} c_k d_{kl} \chi_l \, ,$$

if

$$\phi_k = \sum_l d_{kl} \chi_l \, .$$

Consequently we can predict that we will get $N \mid \sum_k c_k d_{kl} \mid^2$ times the value β_l by measuring quantity B.

If, on the other hand, we had a mixture, that means an ensemble consisting of subensembles, each one of which was again a pure case, then, e.g., $N \mid c_1 \mid^2$ systems of this ensemble have the value α_1 and the state ϕ_1; $n \mid c_2 \mid^2$ systems of the ensemble have value α_2 and the state ϕ_2, etc. Hence we can predict that in future measurements we will get $N \mid c_1 \mid^2 \mid d_{11} \mid^2 + N \mid c_2 \mid^2 \mid d_{21} \mid^2 + \ldots$ times, or $N \sum_k \mid c_k \mid^2 \mid d_{kl} \mid^2$ times the value β_l. (The probability for ϕ_1 is $N \mid c_1 \mid^2$; that of β_l as well as ϕ_1 is $N \mid c_1 \mid^2 \mid d_{11} \mid^2$ and so on for all other states ϕ_k.)

From this it follows that predictions for a pure case differ from those of a mixture, because

$$N \mid \sum_k c_k d_{kl} \mid^2 \text{ and } N \sum_k \mid c_k \mid^2 \mid d_{kl} \mid^2$$

are not generally the same.

20. If $\overline{U}$ is the expectation-value of the quantity U, then

$$\overline{U} = \Sigma \ p_i u_i \ ,$$

where p_i is the probability of the occurrence of u_i. But we can also write

$$\overline{U} = \mathrm{Tr}(PU) \ ,$$

where P is the density matrix and U the matrix of the operator U in a given basic system $\{\phi_n\}$. (Here the ϕ_n are not to be viewed as the eigenfunctions of U.) In a dispersion-free ensemble, every element would have the same value u_k. Therefore $\overline{U} = u_k$, and consequently $(\overline{U})^2 = u_k^2 = \overline{U^2}$, as well as $\mathrm{Tr}(PU^2) = |\ \mathrm{Tr}(PU)\ |^2$.

Now, if we suppose that U is an operator projected on the subspace spanned by the eigenvector ϕ_m, then, owing to the idempotency of U, we also get

$$\mathrm{Tr}(PU) = |\ \mathrm{Tr}(PU)\ |^2 \ ;$$

and since in the present case $\mathrm{Tr}(PU) = P_{mm}$ (P_{ik} being the elements of the density matrix P), then, by means of the fact that $\mathrm{Tr}(P) = 1$ and $\mathrm{Tr}(PU) = 1$, we find that for all i

$$P_{ii} = P_{ii}^2 = 0 \text{ or } 1 \ .$$

But this result is incompatible with the fact that for all possible orthogonal decompositions of a state-function Ψ into $\Psi = \sum_i c_i \phi_i$ in Hilbert space, the condition $\int |\ \Psi\ |^2 dr = \sum_i |\ c_i\ |^2 = 1$ holds, and consequently, $\sum_i P_{ii} = \sum_i c_i^* c_i = 1$. From this it follows that representations of P can always be found for which $P_{11} + P_{22} \neq 1$ and $P_{11} + P_{22} \neq 0$.

21. D. Bohm, "A Suggested Interpretation of the Quantum Theory in Terms of 'Hidden' Variables," *Physical Review*, vol. 85 (1952), pp. 166ff., 180ff.; "Proof That Probability Density Approaches $(\Psi)^2$ in Causal Interpretation of the Quantum Theory," *Physical Review*, vol. 89 (1953), p. 458ff.; "Comments on an Article of Tabakayashi concerning the Formulation of Quantum Mechanics with Classical Pictures," *Progr. Theor. Phys.*, vol. 9 (1953), p. 273ff.; D. Bohm and J. P. Vigier, "Model of the Causal Interpretation of Quantum Theory in Terms of a Fluid with Irregular Fluctuations," *Physical Review*, vol. 96 (1954), p. 208ff.; J. Bub, "Hidden Variables in the Copenhagen Interpretation—A Reconciliation," British Journal for the Philosophy of Science, vol. 19 (1968), pp. 185–210; "What Is a Hidden Variable Theory of Quantum Mechanics?" *Int. J. Theoret. Phys.*, vol. 2 (1969), pp. 101–23.

22. Bub, "Hidden Variables and the Copenhagen Interpretation—A Reconciliation," p. 186.

23. With respect to the intellectual-historical background pertaining to Bohr's physics, cf. M. Jammer, *The Conceptual Development of Quantum Mechanics* (New York, 1966); and Meyer-Abich, *Korrespondenz, Individualität und Komplementarität* (see note 8, this chapter).

24. Bub, "Hidden Variables and the Copenhagen Interpretation—A Reconciliation," p. 206; P. Feyerabend, "Problems of Empiricism," in R. G. Colodny, ed., *Beyond the Edge of Certainty: Essays in Contemporary Science and Philosophy* (New Jersey, 1965).

Chapter 7

1. C. F. von Weizsäcker, *Zum Weltbild der Physik* (Stuttgart, 1958), p. 301.
2. H. Reichenbach, *Philosophische Grundlagen der Quantenmechanik* (Basel, 1949).
3. P. Mittelstaedt, *Philosophical Problems of Modern Physics* (Dordrecht, Holland, 1976).
4. P. Lorenzen, *Meta-Mathematik* (Mannheim, 1962).
5. Mittelstaedt, *Philosophical Problems of Modern Physics*, p. 177.
6. W. Stegmüller, *Theorie und Erfahrung* (Berlin, 1970).
7. P. Suppes, "The Probabilistic Argument for a Non-classical Logic of Quantum Mechanics," *Philosophy of Science*, vol. 33 (1966), pp. 14–21.
8. Stegmüller, *Theorie und Erfahrung*, p. 440.
9. Ibid., p. 452.
10. Ibid., p. 455.
11. In Reichenbach's terminology: A or next A (nächst A) implies next next B.
12. Here we need not go into the works which have appeared on quantum logic by E. Scheibe (*Die kontingenten Aussagen der Physik: Axiomatische Untersuchungen zur Ontologie der klassischen Physik und der Quantentheorie* [Frankfurt a. M, 1964]), H. Lenk (*Kritik der logischen Konstanten,* [Berlin, 1968]), and J. D. Sneed (Quantum Mechanics and Probability Theory," *Synthesis,* part I, vol. 21, 1970), since I have limited myself here to only those authors who maintain a view of the *incompatibility* of quantum theory and classical logic.

Chapter 8

1. It will probably occur to the reader that I speak first of facts (*Tatsachen*) and then of factual assertions (*Tatsachenbehauptungen*). But if the latter are accordingly dependent on theories, then the former cannot be something absolute, since the content of a fact is scientifically only given in terms of an assertion. If I assert: "This current has 100 amps," then I express a fact. If this assertion is then dependent on theories—and this is without doubt the case—then the fact, which is intended as the object of the assertion, is also dependent on theories.
2. I first introduced the concept "historical situation" in my article, "Philosophische Fragen der Zukunftsforschung," *Studium Generale,* vol. 24 (1971).
3. Cf. chapter 5.
4. Cf. chapter 6.
5. W. P. Webb, "The Historical Seminar: Its Outer Shell and Its Inner Spirit," *Mississippi Valley Historical Review,* vol. 42 (1955/56).
6. Cf. chapter 13.
7. Cf. chapter 5.
8. Following this brief sketch outlining my opposition to Hegel, I might also add a short remark concerning Marx. When Marx attempts to represent historical processes as dependent upon productive forces in the final analysis, to my way of thinking this means that the exact same element of the system-ensemble becomes the basis of all movement. But in doing this Marx foisted a structure onto history as a whole that had been extrapolated from a description of a system which pertained merely to one particular epoch—the so-called First Industrial Revolution—and one which was perhaps only partially adequate even to this. Such a view is representative of an utterly nonhistorical monism.
9. The meaninglessness of absolute statements concerning the nature of space results from the fact that all measurement results which might be used to research this can always be interpreted as either mirroring the geometry of space or merely following from the determinant structures of that physics under which the results

were produced. For example, in antiquity we find a spacial geometry of the universe developed in terms of Aristotelian natural philosophy and deviating from the Euclidean conception. Conversely, Descartes developed his physics by assuming the Euclidean nature of space and, as already noted, by grounding this assumption in his rationalistic philosophy. Finally, Einstein set out once again on the basis of physics—i.e. from the principle of the covariance of all coordinate systems—when he interpreted the universe in terms of Riemann space.

10. For an introduction to this concept, see again "Philosophische Fragen der Zukunftsforschung."

11. Th. Kuhn, *The Structure of Scientific Revolutions* (Chicago, 1964).

12. Thucydides was probably the first to grasp this kind of decay in a somewhat intuitive manner when he saw that the real disaster of his age consisted in the ruinous confusion which marked the final passing of the old Homeric harmony.

Chapter 9

1. R. Descartes, *Principia Philosophiae, Pars Secunda,* XLVII, in *Oeuvres,* ed. C. Adam and P. Tannery, vol. 8.

2. Ibid., XLIX.

3. Ibid., LII. There it reads: "Nec ista egent probatione, quia per se sund manifesta." The French text is even more revealing: "Et les demonstrations de tout cecy sont si certaines, qu'encore que l'experience nous sembleroit faire voir le contraire, nous serions néantmoins obligez d'adjouster plus de foy à nostre raison qu'à nos sens." *Oeuvres,* vol. 9.

4. "Hic vero diligenter advertendum est, in quo consistat vis cuiusque corporis ad agendum in aliud, vel ad actioni alterius resistendum." *Principia,* Pars Secunda, XLIII.

5. "Visque illa debet aestimari tum a magnitudine corporis in quo est, et superficiei secundum quam istud corpus ab alio disiungitur; tum a celeritate motus, ac natura et contrarietate modi, quo diversa corpora sibi mutuo occurrunt." Ibid., XLIII.

6. "Alia autem sunt in rebus ipsis, quarum attributa vel modi esse dicuntur; alia vero in nostra tantum cogitatione. Ita, cum tempus a duratione generaliter sumpta disinguimus, dicimusque esse numerum motus, est tantum modus cogitandi." *Principia,* Pars Prima, LVII. The French text is even more precise: "De ces qualitez ou attributs, il y en a quelques—uns qui sont dans les choses mesmes, et d'autres qui ne sont qu'en nostre pensée." *Oeuvres,* vol. 9.

7. "Sed ut rerum omnium durationem metiamur, comparamus illam cum duratione motuum illorum maximorum, et maxime aequabilium, a quibus fiunt anni et dies; hancque durationem tempus vocamus. Quod proinde nihil, praeter modum cogitandi, durationi generaliter sumptae superaddit." *Principia,* Pars Prima, LVII. Here again the French text offers us an informative supplement, where it reads "bien qu'en effet ce que nous nommon ainsi ne soit rien, hors de la véritable durée des choses, qu'une facon de penser." *Oeuvres,* vol. 9.

8. "Sed si non tam ex vulgi usu, quam ex rei veritate, consideremus quid per motum debeat intelligi, ut aliqua ei determinata natura tribuatur: dicere possumus *esse translationem unius partis materiae, sive unius corporis, ex vicina eorum corporum, qua illud immediate contingunt et tanquam quiescentia spectantur, in viciniam aliorum.*" *Principia,* Pars Secunda, XXV.

9. "Addidi denique, translationem illam fieri ex vicinia, non quorumlibet corporum contingorum, sed *eorum* duntaxat, quae *tanquam quiescentia spectantur.* Ipsa enim translatio est reciproca, nec potest intelligi corpus AB transferri ex vicinia corporis CD, quin simul etiam intelligatur corpus CD transferri ex vicinia corporis

AB: ac plane eadem vis et actio requiritur ex una parte atque ex altera." Ibid., XXIX.

10. The italics are mine; the Latin text runs as follows: "Intelligimus etiam perfectionem esse in Deo, non solum quod in se ipso sit immutabilis, sed etiam quod modo quam maxime constanti et immutabili operetur: adeo ut, iis, mutationibus exceptis, quas evidens experientia vel divina revelatio certas reddit, quasque sine ulla in creatore mutatione fieri percipimus aut credimus, nullas alias in eius operibus supponere debeamus, ne qua inde inconstantia in ipso arguatur. Unde sequitur quam maxime rationi esse consentaneum, ut putemus ex hoc solo, quod Deus diversimode moverit partes materiae, cum primum illas creavit, iamque totam istam materiam conservet eodem plane modo eademque ratione qua prius creavit, eum etiam tantundem motus in ipsa semper conservare." Ibid., XXXVI.

11. Mouy writes: "Il y a là une erreur, parce que cette manière de considérer le mouvement est en désaccord complet avec la relativité que Descartes lui avait attribuée en principe. Si le mouvement est relatif, sa 'détermination' n'est pas une propriété absolue qui être considérée à part et qu'on ait, par exemple le droit d'inverser." ("There is an error in this, for this manner of assessing movement is in complete contradiction to the relativity that Descartes in fact attributed to it. If the movement is relative, its 'determination' is not an absolute property which can be separately considered, which, for example, we are permitted to reverse.") *Le développement de la physique cartésienne* (Paris, 1934), p. 22. Koyré remarks: "En effet, ce n'est pas seulement avec les lois du choc que la relativité cinétique du mouvement se révèle incompatible. Elle l'est déjà avec celle de la conversation du mouvement, comprise, comme Descartes veut expressément la comprendre, comme conversation de la *quantité* de mouvement; car il est évident que si l'on attribue—à quoi la réciprocité et la relativité cinétique nous donneraient le droit—*la même vitesse* tantôt au grand, tantôt au petit corps qui se rapprochent ou s'éloignent l'un de l'autre, on obtiendra des *quantités de mouvement* très différentes. Or, on ne peut admettre que Descartes soit resté insensible à des contradictions aussi flagrantes; ni qu'elles lui aient échappé." ("In effect, it is not merely the case that the kinetic relativity of movement reveals itself as incompatible with the laws of impact, but actually that it is already incompatible with the conservation of movement, which Descartes expressly wished to understand as consisting in the conservation of the *quantity* of movement. It is evident that if *the same velocity* is attributed on the one hand to large, and on the other to small, bodies, on both their approach or their withdrawal [after the impact] from one another—as the reciprocity and kinetic relativity would permit us to say— we would in fact obtain quite different *quantities of movement*. But we cannot suppose that Descartes remained insensitive to such flagrant contradictions, nor that they simply escaped his notice.") Koyré adds to this: "L'ultrarelativisme de sa notion du mouvement n'est pas originel chez Descartes. Il ne l'adopte, croyons-nous, que pour pouvoir concilier l'astronomie copernicienne, ou, plus simplement, la mobilité de la terre, visiblement impliquée par sa physique . . . , avec la doctrine officielle de l'Eglise. Effort qui n'aboutit qu'à rendre la mécanique cartésienne contradictoire et obscure." ("The ultrarelativism of his notion of movement is not genuinely Cartesian. He adopted this notion, in our opinion, in order to reconcile Copernican astronomy, or more simply the movement of the Earth evidently implied by his physics, with the official doctrine of the Church. The result of this effort is that Cartesian mechanics is rendered contradictory and obscure.") *Galilé et la Loi d'inertie* (Paris, 1939), p. 329.

12. "Inventis iam quibusdam principiis rerum materialium quae non a praeiudiciis sensum, sed a lumine rationis its petita sunt, ut de ipsorum veritate dubitare

nequeamus, examinandum est, an ex iis solis omnia naturae phaenomena possimus explicare." *Principia,* Pars Tertia, I.

13. R. Descartes, *Discourse on Method, Optics, Geometry, and Meteorology,* tr. Paul J. Olscamp (Indianapolis and New York: Library of Liberal Arts, 1965), p. 50. [Hübner cites the German translation by L. Gäbe (Hamburg, 1960), p. 101; and he writes here:] In distinction to the previous quotations, here and in what follows I do not present the original text, since we are more concerned with the general context than the exact literal sense of the passages.

14. R. Descartes, *Meditations and Selections from the Principles of René Descartes,* tr. John Veitch (La Salle, Ill.: Open Court, 1950), pp. 209–11. [Hübner's reference is to the German translation by A. Buchenau (Hamburg, 1955).]

15. Cf. the passage as quoted by Mouy, *Le développement de la physique cartésienne,* p. 193.

16. Quotation taken from Mouy, ibid., p. 193.

17. "Or, il est remarquable que, pour parvenir à ce résultat merveilleux, Huygens est parti, tout simplement, des hypothèses cartésiennes . . . Huygens est strictement cartésien dans ses principes, et il est impossible de trouver cas plus authentique de développement des postulats cartésiens." Mouy, *Le développement de la physique cartésienne,* p. 197.

Chapter 10

1. P. A. Schilpp, ed., *Albert Einstein: Philosopher-Scientist,* 2d ed. (New York, 1951), p. 400.

2. Ibid., p. 401.

3. Ibid., p. 400.

4. Ibid., p. 400. Exactly what is to be understood by "reality" is beside the point in this context. Here we are concerned with the *content* of the theory, that is, with the fact that the "investigator's passion" is not satisfied by arbitrary, perhaps even unintelligible, theories churned out at random by computers in whatever quantity; rather, this passion seeks to produce a *picture* of nature, founded in whatever way might be possible at the time.

5: Ibid., p. 400; author's italics.

6. Ibid., p. 400.

7. Ibid., p. 398; author's italics.

8. Ibid., p. 392; author's italics.

9. Ibid., p. 393.

10. Stated more precisely this means that for the points in the four-dimensional space-time manifold for which the spatial coordinates have the same value in different synchronized coordinate systems (shown here with and without a bar over the symbol), that is, for those systems where $\bar{x}_i = x_i$, the following equation holds: $\overline{g_{\mu\nu}}\,(\bar{x},t) = g_{\mu\nu}(x,t)$, where the "$g_{\mu\nu}$" stand for the components of the metric tensors, with $\mu,\nu = 0,1,2,3$.

11. In the cosmological principle, Bondi sees the presuppositions for the fact that the same experimental results are to be obtained under the same conditions everywhere in the universe. For this reason he favors a cosmological principle that is even more exact than the one under discussion here, but which is only of significance for the steady-state theory. Cf. H. Bondi, *Cosmology* (Cambridge, 1961), p. 11ff.

12. This can be explained in the following manner: From the postulate of the cosmic substratum and the cosmological principle, we can infer that $g_{00} = 1$, and all other $g_{\mu\nu}$ with an index of 4 disappear. In this way the line-element $ds^2 = g_{\mu\nu}dx^\mu dx^\nu$ becomes $ds^2 = dt^2 + g_{ik}dx^i dx^k$, where t is the time and i,k = 1,2,3. If, then, in

addition, the geometrical relations are taken to change isotropically and homogeneously along with the time in accordance with the cosmological principle, $g_{ik} = |R(t)|^2 l_{ik}$.

Here R(t) is a time-dependent extensional coefficient that is squared, owing to the previously given quadratic form of the line-element. With an appropriate choice of coordinates, we then arrive at the Robertson-Walker line-element:

$$ds^2 = dt^2 - \frac{[R(t)]^2[(dx^1)^2 + (dx^2)^2 + (dx^3)^2]}{\{1 + \frac{1}{4}k[(x^1)^2 + (x^2)^2 + (x^3)^2]\}^2} \, ,$$

where k is the curvature parameter, which can only have the values -1, 0, and $+1$.

13. This formula reads

$$\left(\frac{dR}{dt}\right)^2 = \frac{C}{R} - k + \frac{c^2}{3}\lambda R^2 \, .$$

Here C is the energy constant, λ the cosmological constant as it occurs in the field equations of the general theory of relativity, and c the speed of light. The different solutions to this equation are dependent upon which of the three possible values (-1, 0, $+1$) is assigned to k and whether λ is greater than, less than, or equal to zero.

14. A. S. Eddington, *Space, Time, and Gravitation* (Cambridge, 1920); J. Jeans, *Physics and Philosophy* (Cambridge, 1942).

15. K. Gödel, "A Remark about the Relationship between Relativity Theory and Idealistic Philosophy," in P. A. Schilpp, ed., *Albert Einstein,* p. 562.

16. Kant, *Critique of Pure Reason,* First Antinomy.

17. Cf. the chapter entitled "Infinity and the Actual" in J. D. North, *The Measure of the Universe* (Oxford, 1965). North also shows how a Cantorian enumerative procedure could be applied to an infinite set of galaxies.

18. Kant, *Critique of Pure Reason,* First Antinomy.

19. St. Augustine, *De Civitate Dei,* XI, 6.

20. Cf. M. Bunge, *The Monist,* 1962, p. 126; also G. J. Whitrow, *British Journal for the Philosophy of Science,* 1954, p. 215; and R. Harré, ibid., 1962, p. 110.

21. Kant, *Critique of Pure Reason,* First Antinomy.

22. Ibid.

23. Ibid., B 38–39, tr. Norman Kemp Smith (New York, 1965), p. 68.

24. Ibid., B 46, tr., p. 75.

25. In Schwarzschild's solution of the Einsteinian field equations, it is assumed that there are certain exceptional reference systems in which the $g_{\mu\nu}$ have Minkowski values at infinity. Thus, as A. Grünbaum has demonstrated in his work, *Philosophical Problems of Space and Time,* 2d enlarged ed. (Dordrecht & Boston, 1973), p. 420, the boundary conditions at infinity play the same role here as Newton's absolute space.

26. R. H. Dicke, "Cosmology, Mach's Principle, and Relativity," in C. Dewitt and B. S. Dewitt, eds., *Relativity, Groups, and Topology* (New York, 1964), pp. 222–36.

27. J. L. Synge, *Relativity: The General Theory* (Amsterdam, 1960).

28. K. R. Popper, *The Logic of Scientific Discovery,* 2d rev. ed. (New York, 1968), p. 31. [*Logik der Forschung,* 2d rev. ed. (Tübingen, 1966), p. 6.] (The page

references to the second English edition also correspond to those in the first English edition, published in 1959 [London: Hutchinson & Co.; New York, Basic Books].)

29. Ibid., pp. 38–39. [Hübner quotes the German edition. There is no substantial difference in meaning between the English and German, but the terminology is quite different. In what follows Hübner refers specifically to some of these terms: in particular, "wissenschaftlich indiskutable" (scientifically nondiscussable) and "historisch-genetisch" (historiogenetic or historical-genetic). Hence, for the sake of clarity, I will reproduce the German text as Hübner quotes it. The italics also belong to Hübner: "*psychologisch* gesehen, ohne einen *wissenschaftlich indiskutablen,* also, wenn man will, '*metaphysischen*' *Glauben* an . . . manchmal *höchst unklare . . . Ideen* wohl gar nicht möglich. . . . Dennoch halten wir es für die *wichtigste Aufgabe* der Erkenntnislogik, einen Begriff der empirischen Wissenschaft anzugeben, der . . . insbesondere auch eine klare Abgrenzung gegenüber diesen *historisch-genetisch* manchmal so förderlichen metaphysischen Bestandteilen gestattet." *Logik der Forschung,* 2d ed. (Tübingen, 1966), p. 13.]

30. Here we begin with the following equation derivable from relativistic cosmology:

$$l_0 = \frac{L_1}{4\pi R_0^2 s_k^2(\omega)(1 + z)^2}$$

l_0 signifies the light energy of a galaxy per unit time, unit area that appears to an observer at time t_0; L_1 is the total energy radiating from the same galaxy at time t_1; $4\pi R^2 s_k^2(\omega)$ expresses the surface area of the wave front pertaining to the spherical wave of that light at time point t_0 in Riemannian geometry; k is the curvature parameter ($-1, 0, +1$); and $(1 + z)$ is the redshift. Now, according to the Weber-Fechner law, the relation of the perceived brightness (magnitude) m and the radiation energy l is given by the equation $m = -2.5 \log l_0$. By utilizing logarithms in the original equation given above, we thus arrive at

$$m = M + 5 \log R_0 + 2.5 \log 4\pi + 5 \log s_k(\omega) + 5 \log(1 + z),$$

where $M = -2.5 \log L_1$. Hence here it is a matter of producing a relation between the observable m and the observable z whereby the argument ω of s_k is in turn dependent upon z.

31. The relation between the observable m and the observable z in note 30 of this chapter can only be obtained by means of the given logarithmic equation if (among other things) M is viewed in a sense as a constant, so that the radiation energy of the galaxy either is the same at all times or exhibits the same time dependency. The other tests for relativistic cosmology—for instance, the density test and the age test—proceed on the basis of similar presuppositions which are grounded in the cosmological principle. Hence the radiation test does not represent an exception in terms of the question of falsification treated here.

32. The a priori status of the principle of the equivalence of all reference systems is quite evident here as well, regardless of the fact that this principle comes about here through the application of the a priori postulate of the simplicity of nature, which belongs to the situational context in which Einstein found himself as described above. Indeed, it can be regarded as an empirical fact that in the neighborhood of Earth no difference can be established between gravitational and inertial acceleration. This rests on the equality of inertial and gravitational mass,

something which has been well proved experimentally. But this experimental fact only *makes possible* the equivalence of reference systems; it *does not found* it. To found this we must presuppose that such a difference disappears in any and every local Lorentz frame, anywhere and at any time in the universe.

33. Kant, *Critique of Pure Reason,* B 455, tr., p. 396.

34. North, *Measure of the Universe,* p. 407, writes: "we have found no evidence of immortality in the natural sciences. None has an absolute and permanent value. The individual theory of cosmology is neither true nor false: like any other scientific theory, it is merely an instrument of what passes for understanding."

35. I wish to thank Professor Volker Weidermann, Institut für theoretische Physik, Universität Kiel, for proofing the manuscript of this chapter and for his valuable suggestions.

Chapter 11

1. In regard to Popper's theory of truth, see K. R. Popper, *Conjectures and Refutations: The Growth of Scientific Knowledge* (London, 1965); also, *Objektive Erkenntnis* (Hamburg, 1973).

2. These theses were presented by the LSE philosophers at the symposium sponsored by the Thyssen Foundation (*Thyssenstiftung*) at Schloß Kronberg in July of 1975. Nearly all of the philosophers representative of the theories of Popper and Lakatos at the LSE were present at this meeting.

Chapter 12

1. Cf. W. Stegmüller, *Theorie und Erfahrung* (Berlin, 1970); "Theoriendynamik und logisches Verständnis," in W. Diederich, ed., *Beiträge zur diachronen Wissenschaftstheorie* (Frankfurt a. M., 1974) [and *The Structure and Dynamics of Theories* (New York, 1976)].

2. In this, Stegmüller refers to Kuhn's distinction between normal science and revolutionary science.

Some readers might note the absence here of any critical discussion of Kuhn's theory concerning the structure of scientific revolutions (*The Structure of Scientific Revolutions* [Chicago, 1962]). Without wishing to diminish the historical significance of the role played by Kuhn in the discovery of certain important fundamental problems belonging to the scientific-theoretical realm, we must nevertheless recognize that the new Popperian theory, as well as the theory of Sneed and Stegmüller, have grown out of an insight into the weaknesses implicit in Kuhn's conception, weaknesses which are now rarely disputed. Hence today it is only worthwhile to deal with these later theories. Cf. also my review of Kuhn's book in *Philosophische Rundschau,* 15 (1968), pp. 185–93.

Chapter 13

1. C. G. Hempel, *Aspects of Scientific Explanation* (New York, 1965); P. Gardiner, *The Nature of Historical Explanation* (Oxford, 1961); M. White, *Foundations of Historical Knowledge* (New York, 1969); A. C. Danto, *Analytical Philosophy of History* (Cambridge, 1968).

2. Thus, for example, Herder in his consideration of the philosophy of history regards nations as organisms, von Humboldt compares historical processes with the metamorphosis of plants, Ranke calls various peoples (*Völker*) "Ganzheiten" (wholes or totalities), and Dilthey uses similar expressions; for instance, in his later works he refers to "Bedeutungs-, Wirkungs- und Strukturzusammenhängen" (contexts of meaning, effect, and structure).

3. For example, Herder calls this "Einfühlen" (empathy); Dilthey speaks of "Verstehen" (understanding); Troeltsch calls it "Ahnen" (sensing or intuiting); and Ranke uses the term "Divination" (divining).

4. W. L. Langer, "The Next Assignment," *American Historical Review*, vol. 69 (1963).

5. Cf. W. Stegmüller, Probleme und Resultate der Wissenschaftstheorie und Analytischen Philosophie, 1, *Wissenschaftliche Erklärung und Begründung* (Berlin/Heidelberg/New York, 1969).

6. M. Weber, "Die 'Objektivität' sozialwissenschaftlicher und sozialpolitischer Erkenntnis," in *Soziologie, Weltgeschichtliche Analysen, Politik* (Stuttgart, 1964), p. 234.

7. Here the reader is referred to the interesting work of E. von Savigny entitled "Zur Rolle der deduktiv-axiomatischen Methode in der Rechtswissenschaft," in *Rechtstheorie* (Frankfurt a. M., 1971).

8. A. Schopenhauer, *Die Welt als Wille und Vorstellung, Sämtliche Werke*, vol. 2, ed. A. Hübscher (Leipzig, 1938), p. 505.

9. Ibid., p. 506.

10. E. Gibbon, *History of the Decline and Fall of the Roman Empire*, newly edited by J. E. Bury, 7 vols. (London, 1896–1900).

11. B. G. Niebuhr, *Römische Geschichte*, 3 vols. (Berlin, 1811–32).

12. Th. Mommsen, *Das Römische Staatsrecht* (Berlin, 1887); *Römische Geschichte* (Berlin, 1854–56).

13. M. Rostovtzeff, *Social and Economic History of the Roman Empire* (1926).

14. A Heuss, *Römische Geschichte* (Braunschweig, 1971), p. 575.

15. E. Meyer, *Kaiser Augustus* (Halle, 1924).

16. Cf. also W. Stegmüller, "Der sogenannte 'Zirkel' des Verstehens," in *Natur und Geschichte,* published by the 10th German Congress for Philosophy Kiel, 1972, ed. K. Hübner and A. Menne (Hamburg, 1974).

17. Here I would like to stress that despite the present criticism I believe that I agree with hermeneutical thinkers on some points. But it seems to me that we cannot achieve clarity or even set straight much of what they wish to say until we free ourselves from their murky language. Hence we must apply analytical methods to the historical sciences, methods which the theoreticians of science have hitherto applied only to the natural sciences. In this investigation I hope to be able to demonstrate clearly that this is possible, precisely because both branches of knowledge have the same logical form at their bases (however hidden this might appear at first). Regarding the criticism of hermeneutics, cf. also G. Patzig, "Erklären und Verstehen," *Neue Rundschau*, vol. 3, 1973.

18. J. L. Borges, *Labyrinthe* (München, 1959), p. 89ff. [Cf. *Labyrinths: Selected Stories and Other Writings*, ed. D. A. Yates and E. Irby (New York, 1964), pp. 149 and 155.]

19. Regarding the significance of the Göttingen school, see H. Butterfield, *Man on His Past* (Cambridge, 1969).

20. Cf. E. Cassirer, *Die Philosophie der Aufklärung* (Tübingen, 1932), pp. 269–79.

21. W. P. Webb, "The Historical Seminar: Its Outer Shell and Its Inner Spirit," *Mississippi Valley Historical Review*, vol. 42 (1955/56).

22. Voltaire, "Essai sur les moeurs et l'esprit des nations," *Oeuvres*, XVIII; *Le Pyrrhonisme de l'histoire, Oeuvres*, XXVI.

23. Montesquieu, *De l'esprit des lois*, in *Oeuvres complétes*, vol. 2 (Paris, 1951).

24. Cf. the chapter on Lord Acton in H. Butterfield, *Man on His Past* (Cambridge, 1969).

25. Cf. A. C. Danto, *Analytical Philosophy of History* (Cambridge, 1968).

26. I would like to thank Professor Trunz for pointing out that Goethe expresses similar thoughts rather gracefully in his essay, "Wiederholte Spiegelungen" ("Repeated Reflections"), *Werke* (Hamburg, 1959), Bd. 12, p. 322f. He writes: "The image carried along through time . . . sways . . . back and forth within [us] for many years. What . . . is long retained is finally expressed externally in vivid memory and once again reflected. From this there blooms a desire to actualize everything that might be conjured up from the past. The longing grows, and its satisfaction inevitably demands that we return to the place that we might possess the feeling of it. . . . Here, in the somewhat decayed locality, there now arises the possibility of creating a new present out of the rubble of existence and that which has been handed down to us. . . . If we now consider that repeated reflections not only maintain the past as vivid, but even raise it to a higher life, then we will be reminded of entoptic† phenomena which, in being reflected back and forth from mirror to mirror, do not fade, but rather begin to catch fire; and here we gain a symbol for what has been continuously repeated and is still repeating itself daily in the history of art, science, the Church, and surely also in the history of the political world."

27. W. Schadewaldt, *Die Geschichtsschreibung des Thukydides* (Berlin, 1929).

28. E. Schwartz, *Das Geschichtswerk des Thukydides* (Bonn, 1929).

29. W. Schadewaldt, *Die Geschichtsschreibung des Thukydides* p. 7.

30. Ibid., p. 27.

31. Ibid., p. 24.

32. Cf. K. Hübner, "Philosophische Fragen der Zukunftsforschung," *Studium Generale*, vol. 24 (1971).

33. Cf. "Philosophische Fragen der Zukunftsforschung," ibid.

34. F. von Schiller, "*Sprüche des Konfuzius*," 1st Stanza: "Threefold is the gait of Time: / The Future is drawn hesitantly towards us, / The Now shoots away like an arrow / The Past stands eternally still."

35. Cf., among others, W. Rehm, *Der Untergang Roms im abendländischen Denken* (Leipzig, 1930).

Chapter 14

1. Ktesibios the Alexandrian, aside from making all manner of playful objects, also constructed a water organ and a water pump. Heron of Alexandria developed mechanisms which utilized the pressure produced by compressed air, heated air, and steam. He then put these to use by combining them with cogged wheels, screws, and chambers containing cylinders. A few examples follow: A mechanism by means of which figures brought out libations whenever a burnt offering was presented at the altar; sacrificial vessels which dispensed holy water whenever a gold piece was thrown in; a temple whose doors opened automatically whenever a sacrificial fire was kindled, and closed again as soon as it was extinguished; and a theater composed of automata. See *Heron von Alexandria: Druckwerke und Automatentheater,* Greek and German text by W. Schmidt (Leipzig, 1899).

2. N. Tartaglia, *Quesiti et inventioni diverse* (Venezia, 1564).

3. Cf. K. Hübner, "Von der Intentionalität der modernen Technik," in *Sprache im technischen Zeitalter,* Heft 25, 1968.

†The word entoptic (*Entoptik*) refers to a doctrine of the phenomena of light and color within certain bodies as these phenomena are transferred from mirror to mirror. Goethe was working on this doctrine in connection with his color theory.

4. Nasmyth continues: "[Every bolt and nut] neither possessed nor admitted of any community with its neighbours. To such an extent had this practice been carried that all bolts and their corresponding nuts had to be specially marked as belonging to each other. Any intermixture that occurred between them led to endless trouble and expense, as well as inefficiency and confusion, especially when parts of complex machines had to be taken to pieces for repairs. . . . In his [K. H. Maudslay's] system of screw-cutting machinery, and in his taps and dies, and screw-tackle generally, he set the example, and in fact laid the foundation, of all that has since been done in this most essential branch of machine construction. . . . Mr. Maudslay took pleasure in showing me the right system and method of treating all manner of materials employed in mechanical structures." Quotation taken from F. Klemm, *A History of Western Technology*, tr. D. W. Singer (New York, 1959), pp. 283–84.

5. The following statement by Nasmyth is also worthy of note: "He [Maudslay] loved this sort of work for its own sake, far more than for its pecuniary results." Ibid., p. 285.

6. U. Wendt, *Die Technik als Kulturmacht in sozialer und geistiger Beziehung* (Berlin, 1906).

7. E. Fink, "Technische Bildung als Selbsterkenntnis," in *VDI-Zeitschrift*, Bd. 104 (1962), p. 678f.

8. G. Foerster, *Machtwille und Maschinenwelt* (Potsdam, 1930).

9. F. Bacon, *Novum Organum* (1620); *Works* (London, 1857ff.), vol. 1 (Faks.—Neudruck, Stuttgart-Bad Cannstatt, 1963).

10. F. Dessauer, *Streit um die Technik* (Frankfurt a. M., 1956), p. 215f.

11. K. Marx, *Grundrisse der Kritik der politischen Ökonomie* (Berlin, 1953).

12. E. Jünger, *Der Arbeiter* (Hamburg, 1932).

13. Th. Litt, *Naturwissenschaft und Menschenbildung*, 2d ed. (Heidelberg, 1954).

14. E. Spranger, *Lebensformen*, 3d ed. (Halle, 1922).

15. H. Fischer, *Theorie und Kultur* (Stuttgart, 1958).

16. F. G. Jünger, *Die Perfektion der Technik*, 2d ed. (Frankfurt a. M., 1949).

17. E. von Mayer, *Technik und Kultur: Kulturprobleme der Gegenwart*, Bd. 3 (Berlin, 1906).

18. O. Spengler, *Der Mensch und die Technik* (München, 1931).

19. M. Scheler, *Probleme einer Soziologie des Wissens;* in *Ges. Werke*, Bd. 8 (Bern-München, 1960).

20. M. Heidegger, "Die Frage nach der Technik," in *Reden und Aufsätze* (Pfullingen, 1954).

21. Cf. K. Hübner, "Philosophische Fragen der Zukunftsforschung," in *Studium Generale*, 24 (1971).

22. K. Marx and F. Engels, *Die Deutsche Ideologie*, in *Marx-Engels Werke*, Bd. 3 (Berlin, 1969).

23. Dessauer, *Streit um die Technik*, pp. 140–42.

24. E. Kapp, *Grundlinien einer Philosophie der Technik* (Braunschweig, 1877).

25. A. du Bois-Reymond, *Erfindung und Erfinder* (Berlin, 1906).

26. E. Mach, *Kultur und Mechanik* (Stuttgard, 1915).

27. Spengler, *Der Mensch und die Technik.*

28. E. Diesel, *Das Phänomen der Technik* (Berlin-Leipzig, 1939).

29. J. Ortega y Gasset, *Betrachtungen über die Technik* (Stuttgart, 1949).

30. Heidegger, "Die Frage nach der Technik."

31. Cf. W. Stegmüller, *Probleme und Resultate der Wissenschaftstheorie und Analytischen Philosophie*, Bd. IV, *Personelle und statistische Wahrscheinlichkeit*, erster und zweiter Halbband (Berlin, 1973).

Chapter 15

1. A synopsis of this issue can be found in the previously cited work I. Lakatos and A. Musgrave, eds., *Criticism and the Growth of Knowledge* (Cambridge, 1970). See also W. Diederich, ed., *Beiträge zur diachronen Wissenschaftstheorie* (Frankfurt a. M., 1974).

2. E. Cassirer, *The Philosophy of Symbolic Forms*, vol. 2, *Mythical Thought*, tr. Ralph Manheim (New Haven: Yale University Press, 1955). [*Philosophie der symbolischen Formen, Zweiter Teil, Das mythische Denken* (Darmstadt, 1953).] Cassirer, to whom the following exposition owes a great deal, was probably the first, and up till now the last, person to utilize the Kantian categories and forms of intuition as a guiding clue in the examination of mythical structures. Inspired by him, I have employed a similar guide in the following treatment. But this similarity is somewhat superficial because I proceed on the basis of a different standpoint than that of Cassirer, who is a Kantian; thus, as will become evident, the categories and forms of intuition also have a completely different sense.

3. One can follow the course of this process with particular clarity in the writings of Plato: The gods evaporate into the absolute otherworldliness of the ideas, while art is rejected precisely because of its sensuous involvement with the world, which allows it to depict nothing but mere appearance. In the *Republic* this finds expression in the statement that "the poets deceive" (377d–e).

4. Claude Lévi-Strauss, whose investigations into the nature of mythical thought are based on the study of South American and Australian cultures, etc., instead of Greek culture, summarizes his findings under the general heading of "pensée sauvage." In the book by the same title, he writes: "there are still zones in which savage thought, like savage species, is relatively protected. This is the case of art, to which our civilization accords the status of a national park, with all the advantages and inconveniences attending so artificial a formula; and it is particularly the case of so many as yet 'uncleared' sectors of social life, where, through indifference or inability, and most often without our knowing why, primitive thought continues to flourish." (*The Savage Mind*, tr. George Weidenfeld and Nicolson, Ltd. [Chicago, 1966], p. 219.) Hence Lévi-Strauss bases his work on universal mythical *structures*, which in fact show an extensive agreement with one another all over the world, and therefore are not limited to the sphere of Greek myth alone. There are two reasons why the treatment which follows here does limit itself to the realm of Greek myth: First, this offers us a particularly familiar example; and second, as we have already stated, science developed out of *Greek myth* by means of a critical confrontation with the latter. Here we are particularly interested in the investigation of *this* relation.

5. These conceptions are still at work with considerable force in Plato's writings: Eros inspires Socrates to speak (*Phaedrus*, 236–37, 244a); the lover can be "possessed by God" (ibid., 249c–d), etc.

6. In the *Phaedrus*, Plato mentions that every god has his own particular realm, in which he rules (247a). Such conceptions of order obviously belong to the structure of mythical thought. Lévi-Strauss writes in *The Savage Mind* (p. 10): "A native thinker makes the penetrating comment that 'All sacred things must have their place.' . . . It could even be said that being in their place is what makes them sacred, for if they were taken out of their place, even in thought, the entire order of the universe would be destroyed. Sacred objects therefore contribute to the maintenance of order in the universe by occupying the places allocated to them. Examined superficially and from the outside, the refinements of ritual can appear pointless. They are explicable by a concern for what one might call 'micro-

adjustment'—the concern to assign every single creature, object or feature to a place within a class."

7. W. F. Otto, *Die Götter Griechenlands*, 6th ed. (Frankfurt a. M., 1970).

8. The analytic breaking down of the world into abstract, qualitative building blocks was something which remained unknown to myth. This was subsequently introduced by philosophy. Among these "building blocks" we find things like Plato's "Ideas" as well as the notions of the moist, dry, warm, and cold, earth, air, fire, and water, etc. Mythically viewed, however, the constitutive elements of the world are always *structures (Gestalten)*.

9. Herodotus, *The Persion Wars*, tr. G. Rawlinson (New York, 1942), book 2, chapter 53, p. 144. Plato also mentions that Homer was looked upon as the person who was primarily responsible for giving the Greeks their gods and educating them as to their entire form of life ($\pi\acute{\alpha}\nu\tau\alpha$ $\tau\acute{o}\nu$ $\alpha\mathring{\upsilon}\tauo\mathring{\upsilon}$ $\beta\acute{\iota}o\nu$—*Republic*, 606e).

10. I spoke previously of myth as a "closed" historical structure. It is now time to explain this more precisely in order to avoid misunderstanding in this matter. Obviously I do not mean by this that myth, in the form in which it was handed down from Homer and Hesiod, exercised some kind of strict or rigorous control over *all* the Greeks of the mythical age or, for that matter, that these two figures can simply be treated together as if representative of some kind of unified and unitary doctrine. Nowhere do I contest the fact that with both of these figures we can already observe certain signs and symptoms of the decline of myth (especially in Hesiod). All of what has already been said, as well as that which is to follow, bears only on certain *structural elements* which are characteristic traits of myth—and Homer, Hesiod, and Pindar serve simply as examples of this. If, in addition, while doing this, I might perhaps have been guilty of the sin of oversimplifying, generalizing, and idealizing in certain places, I hope that the classical philologists will excuse me and enlighten me as to my errors. However, they should also be aware that a beginning can never be made without having recourse to such pardonable sins, and that the specifics of a subject can never come to light or be collected and arranged in any way at all if one does not initially proceed by introducing certain general categories, which might then be corrected at a later date by means of a kind of feedback action.

11. Lévi-Strauss remarks: "The mistake of Mannhardt and the Naturalist School was to think that natural phenomena are *what* myths seek to explain, when they are rather the *medium through which* myths try to explain facts which are themselves not of a natural but a logical order." (*The Savage Mind*, p. 95.) We might add, however, that these "logical realities" are nothing other than the a priori conceptions of order belonging to myth.

12. G. Krüger comes to a similar conclusion. In his book on Plato, *Einsicht und Leidenschaft* (Frankfurt a. M., 1947), he writes: "What Kant showed with respect to modern scientific experience (*von der modernen wissenschaftlichen Empirie*) holds in general for all experience whatsoever: it is not sense perception alone, but rather over and above this a comprehension and understanding of the givens within the framework of a priori possibilities. However, experience takes on a *religious* character where the essential ground of all mutability and possibility is not to be found in the sovereign ego, but rather outside of it. The personality of the active powers external to us stands in inverse relation to the fundamental consciousness of freedom within us" (p. 14). Thus Krüger goes far beyond the categorial differences between the mythical and scientific modes of thought treated here and attempts to relate both of these to a single primordial ground—to man's own self-conception (*Selbstverständnis*). For Krüger, this a priori self-conception

of man as the absolute condition of his "possible experience" is mythical in that it is characterized by the awareness of being ruled by external powers, whereas it is scientific in that it is conditioned by the "act of spontaneity and freedom" which, like transcendental apperception, is internal to man as something belonging to his essential nature. Krüger also holds (p. 23) that: "The *mythical* mode of understanding results in a view of the world which, though highly peculiar to us today, is nevertheless still an *empirical view of the world*." Finally he writes (p. 24): "The unbounded . . . receptivity," and here is meant that which belongs to mythical man, "allows everything of an overpowering nature to be seen a priori as something which . . . 'personally' exerts control." This is perhaps the place to reiterate my argument against a typical objection which is often raised against this mode of interpretation, a mode I share with Krüger to a large degree (cf. the discussion in note 10, this chapter). We find an example of this type of objection in E. M. Manasse's critical review of Krüger's book (*Philosophische Rundschau*, 1957, Beiheft 1). There he writes: "However, since Krüger's analysis is also intended to point up the differences between the ancient and modern modes of thought, it rests upon . . . rather questionable simplifications. A depiction of mythical reality as something which was so completely devoid of self-consciousness, like that given by Krüger, could only be said to apply to an abstract conception of the former." Now, this may indeed be as true as, for instance, the assertion that not everyone in the Middle Ages was, as a rule, a believer in Christianity, to the extent that the Middle Ages *as a whole* is viewed in this light; but this nonetheless alters nothing with respect to the need and the *right* to work out the universal and fundamental structures of a particular mode of thought in such a way as to be able to differentiate this mode in all clarity from other such modes. Where this right is denied, a danger looms large for any kind of cultural history—namely, the danger that it will become submerged in philological detail and hence unproductive. The present-day demise of classical philology, which, for example, can now claim only a shadow of that esteem and importance that it once enjoyed in the cultural realm, is largely reducible to this deficiency, this loss of imagination and vision.

13. Plato describes this with particular clarity in *Ion* (533d–e), where he speaks of poetic inspiration as a divine power *in* men, and likens this to the power of a magnet because it transmits its power from the Muse to the poet and from the rhapsode to the audience.

14. Having read the manuscript of the present chapter, P. Feyerabend called my attention to the fact that there were mythical periods when a god could only be sought at *one* location, at that place where he dwelt. The idea that a god dwells at one particular place, for instance, at Delphi or on Mount Olympus, is a belief which was also held in later times, even when the god was accorded much greater freedom of movement. However, in this I see no contradiction to the relation of whole and part, substance and person, described here, since the presence of the god could be experienced in many different ways and degrees: sometimes the divinity was merely felt or sensed in a vague manner, while at other times he (or she) was seen directly. A god might then be active everywhere, while on the other hand the complete expression of his (or her) presence was something which could perhaps only be experienced at particular sacred places.

15. V. Grønbech, *Götter und Menschen, Griechische Geistesgeschichte II* (Reinbeck bei Hamburg, 1967). Here I have sought to give the concept of the arche a somewhat more precise meaning than we find in Grønbech's work.

16. Cassirer employs this concept in a similar context (cf. note 2, this chapter). Concerning this we also find a pertinent remark in Krüger's book, *Einsicht und*

Leidenschaft (p. 166): "Whereas the modern view of time, determined by New-tonian physics, considers time as something intrinsically (*an sich*) 'empty' and independent—an object of 'pure intuition,' as Kant formulates it—ancient thought understood it concretely as *the time of a substantial being (die Zeit eines Seienden)*, that is, primarily as the duration and passing away of a living entity, for whom it was its *lifetime*. Thus the word for eternity, *Aion*, also originally meant 'lifetime.' Newton's 'absolute time,' which determines the popular depiction of the concept of time today, is, then, properly speaking, the abstract conception of all possible concrete 'times,' when these are taken in the ancient sense."

17. It is apparent that no allusion to the Newtonian concept of time like that given by Krüger (cf. the previous note) is required in order to clarify the difference between the ancient and modern modes of intuition. The conception of time as a continuum of points *within* which events are ordered according to causal laws, and which *accordingly* makes time into something which seems completely di-vorced from objects, can also be found, for instance, in the relativity theory. Here the only difference is that "empty" time, as something which is in no way func-tionally related to matter in motion, has disappeared.

18. H. Fränkel, "Die Zeitauffassung in der frühgeschichtlichen Literatur," in Tietze, ed., *Wege und Formen frühgeschichtlichen Denkens* (München, 1955).

19. Here again it seems that we are dealing with a universal structure which occurs in all mythically determined cultures. Thus M. Eliade, next to Lévi-Strauss probably the most important contemporary researcher of non-European myth, writes the following: "As a summary formula we might say that by 'living' the myths one emerges from profane, chronological time and enters a time that is of a different quality, a 'sacred' time at once primordial and indefinitely recoverable." (M. Eliade, *Myth and Reality* [New York, 1968], p. 18.)

20. Cf. again Eliade, *Myth and Reality* (p. 8): "the foremost function of myth is to reveal the exemplary models for all human rites and all significant human activities—diet or marriage, work or education, art or wisdom." And (ibid., p. 18): "this is why myths constitute the paradigms for all significant human acts." Admittedly, the expressions "model" and "paradigm" are misleading because here it is not a matter of imitating such structures, but rather of actually reap-propriating and repeating the primordial events. However, that Eliade actually intends to convey this meaning through the use of these expressions can be gathered from the general context of his investigations—something which will become evident later on when we cite some other passages.

21. Cassirer, *Mythical Thought*, p. 39. [German, p. 52.]

22. Ibid., p. 38. [German, p. 51.]

23. Cf. Eliade, *Myth and Reality* (p. 19): "'Living' a myth, then, implies a genuinely 'religious' experience, since it differs from the ordinary experience of everyday life. The 'religiousness' of this experience is due to the fact that one re-enacts fabulous, exalting, significant events; one again witnesses the creative deeds of the Supernaturals; one ceases to exist in the everyday world and enters a transfigured, auroral world impregnated with the Supernaturals' presence. What is involved is not a commemoration of mythical events but a reiteration of them. The protagonists of the myth are made present, one becomes their contemporary. This also implies that one is no longer living in chronological time, but in the primordial Time, the Time when the event *first took place*. . . . To reexperience that time, to re-enact it as often as possible, to witness again the spectacle of the divine works, to meet with the Supernaturals and relearn their creative lesson is the desire that runs like a pattern through all the ritual reiterations of myths."

24. Lévi-Strauss has also depicted the mythical conception of time, which deviates so drastically from our own, in an example concerned with the rites of mourning and the historical or commemorative rites among Australian aborigines: "It can thus be seen that the function of the system of ritual is to overcome and integrate three oppositions: that of diachrony and synchrony; that of the periodic or non-periodic features which either may exhibit; and, finally, within diachrony, that of reversible and irreversible time, for, although present and past are theoretically distinct, the historical rites bring the past into the present and the rites of mourning the present into the past, and the two processes are not equivalent: mythical heroes can truly be said to return." (*The Savage Mind*, p. 237.) "The commemorative and funeral rites postulate that the passage between past and present is possible." (Ibid.) Here one might also note that the rite of the Catholic mass is to be understood mythically insofar as there the original event of the Eucharist is *actually* reenacted, and not merely "commemorated." Concerning the relation of Catholic ritual and ancient sacrifice, see H. Lietzmann, *Messe und Herrenmahl* (Bonn, 1926). And again we might also cite Eliade, *Myth and Reality* (p. 13): "It is here that we find the greatest difference between the man of archaic societies and modern man: the irreversibility of events, which is the characteristic trait of History for the latter, is not a fact to the former."

25. It is also precisely for this reason that for the person imbued with mythical consciousness eternal recurrences could not be viewed as fixed and lifeless events, as they might appear to us today, oriented as we are to the incessant unrest of progress. Everything receives its justification for such a person, as Lévi-Strauss notes, in the simple statement that it has been handed down to him from his ancestors (*The Savage Mind*, p. 236). In reporting on the Australian aborigines, Lévi-Strauss quotes Strehlow's account: "In his myths we see the native at his daily task of hunting, fishing, gathering vegetable food, cooking, and fashioning his implements. All occupations originated with the totemic ancestors; and here, too, the native follows tradition blindly: he clings to the primitive weapons used by his forefathers, and no thought of improving them ever enters his mind." (Ibid., p. 235.)

26. Grønbech, *Götter und Menschen*, p. 169.

27. Ibid.

28. Ibid., p. 170.

29. Grønbech, *Götter und Menschen*, p. 170. Krüger also remarks (*Einsicht und Leidenschaft*, p. 38): "In the world of myth *everything* has another face than that which appears in the later, rationally ordered world." (And by this he obviously does not mean that the ancient world was somehow "irrational," but rather only that it was not controlled by *scientific* reason.)

30. I will dedicate a later publication exclusively to the theory of myth.

INDEX OF NAMES

Gibbon, E., 185, 195, 202
Gilbert, W., 39, 56
Gödel, K., 143, 262 n. 15
Goethe, J. W. von, 266 n. 26
Goodman, N., 162–64
Grønbech, V., 236, 243, 270 n. 15
Grünbaum, A., 247 n. 3, 262 n. 25

Harré, R., 262 n. 20
Hawkings, S. W., 138
Hegel, G. W. F., 86, 113–14, 181, 229
Heidegger, M., 216, 219–20, 267 nn. 20, 30
Heisenberg, W., 13, 15, 17, 18, 19–20, 101, 247 nn. 1, 5, 9, 11
Hekataios, 241
Hellanikos, 241
Hempel, C. G., 174, 264 n. 1 (ch. 13)
Herder, J. G., 174, 264–65 nn. 2, 3
Herodotus, 233
Heron of Alexandria, 208, 266 n. 1
Hesiod, 233, 236, 237, 241, 269 n. 10
Heuss, A., 185
Hobbes, Th., 8
Høffding, H., 84
Hohenburg, H. von, 252 n. 8
Homer, 233, 241, 269 n. 10
Hooke, R., 39
Hübner, K., 250 n. 6, 256 n. 12, 265 n. 16, 266 nn. 32 (ch. 13), 3 (ch. 14), 267 n. 21
Hübscher, A., 265 n. 8
Humboldt, W. von, 174, 265 n. 3
Hume, D., 4–5, 9, 11
Huygens, D., 39, 125, 126, 127, 131, 133–37, 164, 209, 260 n. 11

James, W., 84–85, 88, 111
Jammer, M., 82, 257 n. 23
Jeans, J., 143, 252 n. 14
Jordan, P., 152
Jünger, E., 267 n. 12

Kant, I., 4–11, 23, 44, 86, 88, 89, 105, 109, 114, 138, 139, 143–44, 145–47, 152, 153, 183, 230, 231, 233, 268 n. 2, 269 n. 12
Kapp, E., 219, 267 n. 12
Kepler, J., 39, 42, 51–71, 83, 85, 88, 251 n. 5, 252 nn. 10, 12, 16, 253 n. 22, 254 n. 40
Kierkegaard, S., 84, 85, 88, 111
Kirchner, A., 39
Koyré, A., 130, 251 n. 4, 260 n. 11

Krüger, G., 269–70 n. 12, 270–71 n. 16, 271 n. 17, 272 n. 29
Ktesibios of Alexandria, 208, 266 n. 1
Kuhn, Th., 118, 229, 250 n. 7, 264 n. 2 (ch. 12)

Lachmann, K., 111
Lakatos, I., 63, 64–66, 254 n. 40, 255 n. 41
Langer, W. L., 179
Laue, M. von, 209
Leibniz, G. W., 80, 95, 109
Lenin, V. I., 3
Lenk, H., xii, 258 n. 12
Leonardo da Vinci, 209
LeRoy, E. L., 35
Lévi-Strauss, C., 268 nn. 4, 5, 269 n. 11, 272 nn. 24, 25
Lietzmann, H., 272 n. 24
Litt, Th., 215, 267 n. 13
Lorenzen, P., 95, 96
Lysenko, T., 120–21

Mach, E., 35, 219, 267 n. 26
Machiavelli, N., 201–2
Manasse, E. M., 270 n. 12
Marx, K., 219, 258 n. 8 (ch. 8), 267 n. 11
Maudslay, H., 213, 267 n. 4
Maxwell, G., 255 n. 9
Maxwell, J. C., 120, 139
Mayer, E. von, 267 n. 17
Menne, A., 265 n. 16
Mersenne, M., 39
Meyer, E., 186
Meyer-Abich, K. M., 82, 255 n. 8, 257 n. 23
Mittelstaedt, P., 95–98, 100
Møller, P. M., 84
Mommsen, Th., 185
Montesquieu, Ch. L., 195
More, H., 41
Morgenstern, O., 224
Mouy, P., 130, 134–35, 260 n. 11
Musgrave, S., 253 n. 23

Napoleon, 186
Nasmyth, J., 213, 267 n. 4
Neumann, J. von, 79, 80–81, 224
Newton, I., 30–31, 39, 41, 42, 45, 52, 70, 71, 83, 85, 109, 112, 139, 141, 148, 151, 152, 164–65, 178, 195, 270–71 n. 16
Niebuhr, B. G., 111, 185
Niepce, J. N., 209

GENERAL INDEX

Aberration, constant of, 27–28
Absolute, 19, 111
Absolute convergence, 166
Acceleration, 42; gravitational, 263 n. 32; inertial, 263 n. 32
Action, practical, 68, 226
Action by contact, laws of [*Nahwirkungsgesetze*], 14, 80
Adequacy of definition, criterion, 95
Antinomy: Kant's first, 143; mathematical, 229
Aphelion, 55
Appearance, Kantian realm of, 9
A priori, 6–8, 11, 21–23, 88, 91, 124, 138, 143–44, 150–53, 157, 168–69, 171–72, 188–89, 192, 196, 203, 234, 243, 263–64 n. 32, 269 n. 11
Apsis line, 54, 58
Arche, 236–39, 241–43, 270–71 nn. 16, 17, 272 n. 24
Archimedean theorem, 56–57
Argument domain, 166
Aristotelian: metaphysics, 251 n. 4; physics, 44, 52, 56, 62, 65, 112, 113. *See also* Aristotelianism
Aristotelianism, 38–39, 40, 110, 251 n. 13; "oikeios topos," 39
Art, 4, 8, 10–12, 207, 215, 217, 229–30; Aristotelian concept of form, 10; Greek, 241; mere semblance of the beautiful, 230; Platonic "Idea," 10. *See also* Object, of art
Axiomatic system, 108–9

Basic statements, 25–27, 29–30, 33, 47, 63–64, 139–40, 159, 161
Belief (faith), 9, 149; in facts, 25
Biblical criticism and hermeneutics, 111, 194–95
Boundary between natural and human (cultural) sciences, 49–50
Boundary conditions [*Randbedingungen*], 25, 43, 262 n. 25

Capitalism, 224
Carnot's theory of the steam engine, 209
Cartesian: metaphysics, 129; physics, 106, 125–37; tradition, 82—83

Categories, 7, 42–45, 75, 89, 106, 107, 108, 109, 135, 137, 161, 184, 231, 244, 268 n. 2; scientific-theoretical, 42–44
Catholic mass, 272 n. 24
Causality, mythical versus scientific, 231, 232–36, 240, 243–44
Causal principle, 6, 13–24, 47, 74–75, 82, 85, 152; anomalies in, 80; applicability of, 13–16; definition, 13, 14, 17, 22; invalidity, 13; limited form, 16–17, 22; practical methodological postulate, 22; unlimited form, 16–18, 21, 22, 45
Centrifugal force, 148
Concepts of understanding, 231
Confirmation, degree of, 66–69
Constraints [*Nebenbedingungen*], 169
Contingency, historical, 70, 88, 95, 114
Continuity equation, 18
Copenhagen school, 17–18, 18–22, 101
Copernican revolution, 40
Copernican theory, 39, 40–41, 52, 55, 113, 251 n. 4
Cosmology, 138–54, 232, 236; age test, 263 n. 31; "big bang" theory, 142; cosmological formula, 142; cosmological models, 142–48, 152, 153–54; cosmological principle, 75, 138, 141, 150–52, 261 n. 11; density test, 263 n. 31; Minkowski values, 262 n. 25; postulate of the cosmic substratum, 138, 141–42, 151; radiation test, 149–50, 263 n. 31; redshift, 149–50; relativistic, 138–54; steady-state theory, 152, 261 n. 11
Craft, as early form of technology, 208–9, 218
Cultural history, logical sense of, 192
Cultural (human) sciences [*Geisteswissenschaften*], 49–50, 174–203 passim
Curvature parameter, 149, 261–62 n. 12, 263 n. 30
Cybernetics, 210–12, 222–23; automata theory, 212; control processes, 210–12; control theory, 212; feedback processes, 211; game theory, 212–13; homomorphic relations, 212; information theory, 212; isomorphic relations, 212; operands, 210–11, 213; operators, 210–11, 213; purposes (goals) behind transmission

systems, 211; steering processes, 210–11, 212; switching theory, 212; theory of linguistic structures, 212; transmission systems, 210–12
Cyzikos, battle of, 198

Decision theory, 224–28
Definition domain, 166
Demarcation criterion for distinguishing between the scientific and the nonscientific, 195, 231–32, 243–45; Popper's, 149–50
Democracy and technology, 214–15
Demystification of science, 124
De-mythologizing, 214
De-tabooing, 214
Determinism, 82, 83, 109, 231
Dialectic: in Bohr's philosophy, 84–85; Hegelian, 113–14; Kantian, 143, 153
Dispositional properties, 179
Divination, 175, 265 n. 3
Dogmatism in empirical science, 86
Duhem-Quinean problem, 36–39, 140

Eccentricity of orbital focal points, 54, 57–59
Ego, 6, 7–8, 77
Einstein-Podolsky-Rosen paradox, 72–74, 79
Empathy in the cultural (human) sciences, 175, 265 n. 3
Empirical, 4–6, 21–22, 32–34, 114, 148–49, 153, 161–62, 188–89; "empirically significant," 248 n. 13; purely empirical, 32–34, 188–89
Empirical content, 26, 33, 65–66, 107, 139–40, 148–49, 166–70, 188–89
Empiricism, 4–5, 21–22, 105–6, 124, 248 n. 13; Machean, 147
Energy, 31, 74; principle of conservation of, 255 n. 9
Energy-mass equation, 118
Enlightenment, 105–6, 111, 194–95, 214
Enumerative procedure, 144, 262 n. 17
Epicycles, 40, 251 n. 4
Epistemological, 144, 153
Epistemology (theory of knowledge) [*Erkenntnistheorie*], 77, 155–61. *See also* Logic of knowledge; Truth
Equant circle (orbit), 55
Equant point (*punctum aequans*), 54, 55
Error, theory of calculating, 26–27

Essentialism, 123, 223
Eternal return of the same, 146–47
Event (occurrence) [*Ereignis*], 14–15, 16–17; algebra of, 98–100; historical, 206; mythological, 236–37, 239–40; objectivity of, 145; problem of a "first" event, 147
Exactness, 108, 113, 178, 212–14, 215, 216–17, 220–22
Experience, 6–8, 26–27, 33, 42–45, 85–88, 116–17, 125, 127, 130, 136–37, 140, 147, 148–49, 160, 170, 186, 188–89, 192, 196, 202–3, 221, 231, 233–34, 240, 241, 242–44
Experiment, 17, 19, 20, 36, 42–43, 46, 63, 72–74, 86–87, 143; disturbance introduced by, 16, 72–74; *experimentum crucis*, 63, 252 n. 15
Explaining, 174–82; philosophers of, 175–76
Explication. *See* Systems, historical
Extension, coefficient of, 261–62 n. 12
Extensional concept of a whole, 144, 152

Facts, 5, 25–26, 27–29, 32–34, 36, 45, 47, 64–67, 71, 86, 105–7, 110–11, 112–14, 116, 117–18, 121, 123–34, 136, 139, 140, 155, 156–60, 162–65, 186, 188, 189–90, 202–3, 242; core historical, 200–201
Falsifiability, 21–22, 32–34, 63, 133, 148–50, 163
Falsification, 32–34, 37, 116, 133, 142, 148–50, 153, 162–63, 171, 186, 226, 229; criteria of, 37, 44–45, 47, 63–66; degree of, 43, 47, 253 n. 22
Falsificationism, 162–64
Falsity, 116, 140, 153, 155–65, 171
Family resemblance of objects in the sciences, 123, 158
Field equations, 83, 142, 147
Force, 42, 71, 83, 123, 128, 166–67
Freedom, 23–24, 34, 37; realm of human, 214–16
Function, concept of, 44, 89
Futurology [*Zukunftsforschung*], 217–18; convergence procedure, 218; Delphi technique, 218; morphological method, 218; relevance-tree technique, 218; trend extrapolation, 218

Galaxies, 141, 149–50
Genealogies, 241–42

Geodesic lines, 139
Geometry, 44, 86, 141, 146, 152, 250 n. 9;
  Riemannian, 139, 258–59 n. 9, 263 n. 30
God: Copernicus's belief in, 52, 121; Descartes's belief in, 127–33, 136; Einstein's
  belief in, 82–84; Kepler's belief in, 55,
  69, 70; as the numinous in Kant's philosophy, 8–9. *See also* Numinous
Gods and heroes, 232–41; Achilles, 232;
  Aphrodite, 232, 234–35; Apollo, 232,
  234, 236, 239; Athena, 232, 236, 239;
  Chaos, 233, 236, 237; Chronos, 237 (*see
  also* Time); Eros, 232; Gaia (Earth), 233,
  234; Helios (Sun), 232, 237; Hermes,
  232, 236; Leda, 196; Mnemosyne (Memory), 233; Prometheus, 233; Persephone,
  236, 237; Themis, 233; Uranos (Ouranos—Heaven), 233, 234; Zeus, 196. *See
  also* Mythical substances
Goodman's paradox, 162, 163
Gravitational theory, 31, 39–41, 56, 117,
  139, 151, 252 n. 10
Gravity, law of, 71, 120

Hamilton-Jacobian differential equation, 18
Harmony, 5, 84, 119–20
Hegelianism, 113–14, 117
Heidegger's notion of "uncovering" [*Ent-
  bergen*], 220; of "trap" [*Gestell*], 220
Heliocentric, 53, 55, 64, 69–70
Hera, priestesses of (succession used for
  dating events), 242
Hermeneutical circle, 189–90
Hermeneutical philosophers, 265 n. 17
Heroes (mythical). *See* Gods and heroes
Historical, the (historicality), 37–38, 48,
  70–71, 88, 107–16, 154, 161, 172, 176,
  222–23
Historical sciences, 111, 174–203, 231
Historical situation. *See* Situation, histori-
  cal
Historical systems. *See* Systems, historical
Historistic, 35, 36, 42, 107, 116; theory of
  science, 35–50, 105–24, 161–62
History, 115, 161–62, 174–203, 223
History of science, 35–38, 42, 47–49, 70–
  71, 85–87, 88–89, 108, 115–16, 117, 149,
  154, 161–62, 171; critical function of, 48,
  89; as propaedeutic for the theory of science, 45, 47–50
Hooke's law, 169
Homo faber, 219

Horror philosophiae, 88
Human (cultural) sciences, 49–50, 174–203
  passim
Humanism, 40, 52, 55, 112–13, 117, 121
Humanity, 216
Hypotheses, 36, 66–69

Idea, 138, 153–54, 160; Platonic, 10,
  269 n. 8; regulative, 109–10, 156, 158,
  160–61, 183
Ideal [*Ideelles*], 235, 240
Idealism, 76–77, 156, 192
Ideal type of Max Weber, 182–83
Image set, 166
Immanent world-view of myth, 231
Impact, laws of, 125–27, 129–31, 132–36
Indeterminism, 79–82, 231
Induction, 66–69, 162–64, 231
Inductionism, 162–64
Inductive logic, 66–69
Inertia, principle of, 23, 38, 52, 56, 65,
  109, 113, 136, 139
Inertial systems and referent frames, 45,
  118, 120, 139
Inference: by analogy, 253–54 n. 27; direct
  or immediate, 253–54 n. 27; inverse,
  253–54 n. 27; predictive, 67, 253–
  54 n. 27
Influx (influence) of divine substance, 235
Input quantities, 210–11
Intensional concept of a whole, 144, 152
Interphenomena, 14–15, 79–80
Interpolation, 27, 32
Intuition, forms or structures of, 146, 231,
  235–41, 268 n. 2, 270 nn. 16, 17
Inventor, 209, 219

Judgments: categorial, 89; of one-place
  predicates as opposed to predicates of
  more than one place, 89; synthetic a
  priori, 23, 44

Kairos, 235
Kant's first antinomy, 143–47
Kepler: *Astronomia Nova,* 51–71; determination of Earth orbit, 52–53; determination of Mars orbit, 51, 53, 56–58, 65;
  determination of distance of planets
  from Sun, 56–60; extrapolations used in
  generalizing, 55, 70; geocentric longitudes, 53; heliocentric longitudes, 54;
  hypothesis vicaria, 51, 54, 58, 62, 65, 70;

law of the radii, 55, 58; laws, 62, 71; metaphysics, 52, 65; Newton's theories as inductive generalizations from Kepler's, 71; solar mysticism, 55, 69–70; Sun, 55, 58–60, 62, 66, 69–70, 71; theology, 52, 65

Knowledge: absolute, 88, 124; as divine revelation, 127–31; divine versus practical, 131–32, 135–36; hermeneutical problem of, 189–90, 265 n. 17; in the historical sciences, 174–75, 184; scientific versus general, 44. *See also* Logic; Systems; Theories

Lathe-support, development of, 213
Laws: of action by contact [*Nahwirkungsgesetze*], 14, 80; of causality (scientific versus mythical), 231–34, 237, 239 (*see also* Causal principle); of conservation of momentum, 126, 128; of excluded middle (*tertium non datur*), 92, 94, 95, 97; of falling bodies, 4, 6; of gravity, 71, 120 (*see also* Gravitational theory); seven, of historical processes, 111, 115; of impact, 125–27, 129–31, 133–36. *See also* Natural laws
Lever, principle of, 55–56
Logic, 5, 35, 66–69, 79, 90–101, 108, 144, 156, 244; Boolean set theory, 100; dialogical, 95–96; effective propositional, 97, 101; inductive, 66–69; "logic of knowledge," 149; nonclassical, 98, 100; propositional, 95, 97, 100; quantum, 79–80, 90–101; rules of, 94–95; situational, 117, 152; syllogistic, 94–95
Logographers, 241
Logos, 230
Love, historical relativity of concept, 115

Marathon, battle (modes of dating), 237–38
Mass, 31, 42, 71, 83, 89, 106, 126–27, 128–30, 167, 263 n. 32; as a function, 166
Materiality: in Cartesian theory, 128–30; contra ideality, 235
Mathematical: antinomies, 229; models, 108, 211–12
Matrix: of consequences, 224–25; density, 81, 256–57 n. 20; operator, 256–57 n. 20; probability, 225; use, 225
Matter, 41, 141, 143, 146–47; density of, 141, 143
Maxwell's light theory, 120, 139

Measurability, 19–20, 22, 72–74
Measurement: error analysis, 16, 26–27; exactness, 16, 26, 32, 36; instruments, 21, 26, 36, 42, 44, 73–74, 76, 78–79, 209; metric tensors, 142; theory of, 26, 36, 73–74, 146
Measuring, 16–17, 19–20, 27–28, 42–43, 73–74, 76, 78–79, 97, 146, 163, 167, 168
Mechanics, 127–31, 133–37, 141, 151, 164; formalization of classical particle (CPM), 166–67
Mercury (orbit), 62, 119, 164
Metalanguage, 155, 157
Metamorphosis, 232
Metaphysics, 149, 176; metaphysical realism, 156–60. *See also* Aristotelianism; Aristotle; Kepler
Metatheory, 32–34, 157, 173
Methodological considerations, 33, 63, 82, 87, 91
Methodology, 63, 64, 67–68, 71, 82, 86–87
Microphysics, 89, 159. *See also* Einstein-Podolsky-Rosen paradox
Miracles, 9–10
Modus cogitandi, 128, 129, 130
Modus in rebus extensis, 128
Modus in rebus ipsis, 128, 129, 130
Momentum, 73, 77, 80, 83, 94, 100, 106, 128, 130, 131, 141; law of conservation of, 126, 128
Movement, 56, 128–31, 135, 143, 232, 260 n. 11
Mutation. *See* Systems, historical
Mysterium tremendum et fascinosum, 9
Mystery cults, 240
Myth (mythology), 229–45; as immanent world-view, 231
Mythical festivals, 240–41, 243
Mythical substances: Chaos, 233, 237; Custom (use), 234; Day, 233, 236, 237; Death, 234; Delusion (infatuation), 232, 235; Dream, 234; Erebos, 233; Fire, 234; Justice, 234, 235; Love, 235; Night, 233, 234, 236, 237; Order, 233; Seasons, 236, 237; Sleep, 234; Spring, 239; Sun, 234
Mythographers, 241

Narration, 180–81
Narrative sentences, 196–97
Natural laws, 3–5, 6, 7, 9–10, 11–12, 19, 26, 27–29, 31, 33, 43, 67, 73, 101, 161, 163, 174–75, 176–77, 178–81, 182, 195,